Unmaking Botany

Unmaking Botany

Science and Vernacular Knowledge in the Colonial Philippines

KATHLEEN CRUZ GUTIERREZ

Duke University Press *Durham and London* 2025

© 2025 DUKE UNIVERSITY PRESS. All rights reserved
Project Editor: Michael Trudeau
Cover design by Matthew Tauch
Typeset in Garamond Premier Pro by Westchester Publishing Services

Library of Congress Cataloging-in-Publication Data
Names: Gutierrez, Kathleen C., author.
Title: Unmaking botany : science and vernacular knowledge in the colonial Philippines / Kathleen Cruz Gutierrez.
Description: Durham : Duke University Press, 2025. | Includes bibliographical references and index.
Identifiers: LCCN 2024025083 (print)
LCCN 2024025084 (ebook)
ISBN 9781478031482 (paperback)
ISBN 9781478028277 (hardcover)
ISBN 9781478060475 (ebook)
Subjects: LCSH: Botany—Philippines—History. | Botany—Spain—Colonies. | Traditional ecological knowledge—Philippines. | Ethnoscience—Political aspects—Philippines. | United States—Territories and possessions—History. | United States—Foreign relations—1865–1921.|Philippines— Colonization—History.
Classification: LCC QK21.P6 G88 2025 (print) | LCC QK21.P6 (ebook) | DDC 581.9599—DC23/ENG/20241217
LC record available at https:// lccn.loc.gov/2024025083
LC ebook record available at https:// lccn.loc.gov/2024025084

Cover art: Plate 68, "Asclepiadeas," from *Sinopsis de Familias y Generos de Plantas Leñosas de Filipinas*, vol. 2, atlas (Manila: Chofré y Compañía, 1883). Courtesy of Biblioteca Digital of the Real Jardín Botánico.

Support for the publication of this book was provided by the Association for Asian Studies' First Book Subvention Program.

To my parents, ESTRELLA AND HERMES,
for giving me the stars and the gods

Contents

A Note on Orthography, Terms, and Formatting
ix

INTRODUCTION
Sovereign Vernaculars
1

PART I
A Botany at Its Most Defined

1
AN ASYMPTOTIC TAXONOMY
29

2
SCIENTIFIC STATECRAFT
55

PART II
Science in a Place of Flux

3
UBIQUITOUS SAMPAGUITA
85

4
WOVEN TRANSFORMATIONS
107

PART III
Assembling a Wider Expanse

5
FIELD LABOR'S MENACE
135

6
THE LATIN BABBLE
159

CONCLUSION
Of Place, Moment, and Source
183

Acknowledgments	Notes	Bibliography	Index
199	205	235	273

A Note on Orthography, Terms, and Formatting

I do not italicize the proper names of institutions that appear in Spanish or other European languages. Institutional abbreviations follow the Spanish names and not my English translations. For phytonyms or plant names, I include original spellings as they appear in sources. For instance, malacquit and malagcquit both likely refer to the "malagkit" of today. I provide the original Latin spelling for binomials and for the original species name found in my materials even if more contemporary updates have been made to the identification. Such instability is instructive. In the book, I touch on how colonial investigations of plants grappled with changing orthographic or transliterative conventions.

Sources interchange the spelling of Regino García's second surname, Basa (or Baza). I have chosen to use "Basa" to reflect its spelling in the nineteenth-century source material. Since I rely on the orthography that appeared in late nineteenth-century records held in Madrid, I also use "Sebastián Vidal y Soler" instead of the Catalan variant, "Sebastià Vidal i Soler." I do preserve, however, the Catalan variants for other Catalan actors' names that appear in other primary and secondary sources. I provide the birth and death years for historical actors when available, and those years appear only with my first mention of an actor in the book. For all publications and institutions, I include parenthetically the year published or founded at first mention only. Aside from publication titles or purposes of emphasis, Latin is the only language I italicize. I choose to set it apart textually from other languages, such as those of the colonial Philippines including Spanish and English. These languages were and continue to be everyday vernaculars of plant-knowing.

"Filipino" refers to the distinct political identity that emerged through the nineteenth century, whose use accelerated in the final decades of the century to denote people of the Philippine colony. Because of the growth of "Filipino" as an identity in the late nineteenth century, I hesitate to refer to all Philippine-born actors historically as such, especially if I have information on the ethnolinguistic community to which the actor was known or claimed to belong. "Non-Christian" tribal identity became especially marked in the early twentieth century as US anthropology took to the Philippines. I use identity markers, such as Bagobo, as they emerged in archival material, knowing that these actors' presumed ethnic categories were flexible yet formalizing with enhanced colonial ethnological research. I do not use "Filipinx," the gender-neutral term. "Filipino" and "Filipina" were gendered and gendering markers during the period of the present study, and these, as I show, were part of the floral imagination. I use "US" interchangeably as a demonym and as a modifier. Furthermore, I use "Spanish" as a demonym and as a modifier and to refer to Castilian Spanish or the Iberian Spanish language.

For image captions, the language "once known" replaces "unknown" to describe individuals that I have not been able to identify by name. Following the practice of particular Indigenous scholars in the United States and curators in Australia, I use "once known" to recognize that these individuals, though perhaps "unknown" to me or my reading audience, were nevertheless known individuals at some point in time.

Introduction

SOVEREIGN VERNACULARS

> There, crops overflow
> beautiful, thick with offerings;
> there, forests of pure wealth,
> in the river's mud a golden possession,
> hundreds of thousands of scallops offer
> their jewels at the seashore,
> there, along the mountain range,
> absent of predators,
> abundant in wild honey
> —"ANG MUTYÂ NG KASILANGANAN," *Renacimiento Filipino*, 1912

The Philippine environment has engrossed newcomers for centuries. Its charms have beckoned botanists and the botanizing, poets and bureaucrats alike. Today, scientists speculate the archipelago may host the highest concentration of unique species per unit area in the world.[1] They estimate at least 40 percent of its plants to be endemic, growing nowhere beyond the 7,641 islands embraced by hundreds of miles of open sea.[2] But the very condition for marveling at these figures—for marveling at the notion of a Philippine environment—emerged at the nexus of colonization, science, and nature. Foreign observers put in place the idea that an island chain of polylinguistic littoral, riverine, and upland polities could be investigated as a singular colonial environment. In addition to overwhelming plant forms, which such observers studied, cataloged, and classified, that same environment also presented ways of knowing nature that challenged their investigations.

This book is a history of botany in the colonial Philippines, and within this history are ways people knew plants. These ways, from the embodied to the patriotic, the cosmological to the systematic, illuminate the vegetal thoughtworlds present on an archipelago once administered by two successive empires. For both the Spanish and US colonial projects, the science of botany brought order to tropical flora to serve intellectual, commercial, and political interests. Botany's rise as a self-proclaimed international science coincided with the concluding years of the Spanish and the early years of the US colonial regimes, and both deployed similar strategies of botanical systematization.

As this book reveals, even as colonial botanists sought to regiment the Philippine environment along renewed virtues of Linnaean botany, alternative knowledges of nature persisted. I term these "sovereign vernaculars," or insight into plants that made and *un*made the science. Sovereign vernaculars revealed locally nuanced ways by which individuals came to know the plants around them, at times exposing the philosophical unsteadiness, the labor fragility, and the disciplinary limits of botany. Nevertheless, botany's continuance into the early twentieth century came from its encounter with vernaculars. The science's imperial dominance, a sovereignty over how others may come to know plants, thus materialized from its grappling with such divergent insights. The tension present in both categories—of the sovereign, and of the vernacular—drives this book.

The following pages call for a spacious definition of the vernacular, which, in the history of Anglo-European botany, has customarily connoted that which is not Latin or not a Latinized scientific plant name. All plant names—nombres vulgares, local monikers, the common—fell and continue to fall under this category. I reconceptualize the vernacular as more than just the non-Latin and define it instead as expressions of plant knowledge that include and yet are more than names. These expressions emerged both in solitary moments and throughout extensive forest expeditions. They could be heard over a bandurria accompaniment or during a field interview. Sometimes they skirted botanists' gaze or deeply unsettled it. Often, they lived in everyday locations, leaving botanists to puzzle over whether

FIGURE I.I. "Balete tree on Tuai," no. 136, ca. 1910s, Elmer D. Merrill lantern slides, Archives of the New York Botanical Garden. Reproduction permission courtesy of the Archives of the New York Botanical Garden.

they constituted scientific knowledge, its complete opposite, or simply something else. Most significantly, these vernaculars demonstrated expertise that, like the science, remained dynamic, historically contingent, and socially entangled.

This broader definition of the vernacular encourages renewed scrutiny of archival source material and colonial botany tomes. Take, for example, a lantern slide titled "Balete tree on Tuai" (see figure I.1). The name of the black-and-white image calls attention to a balete, or a species of *Ficus*, growing on the commercially profitable bishop wood tree known locally by some as tuai. The image includes a field assistant, likely hired in the locality, whose name does not accompany the photograph. Captured in the first decade or so of US colonial occupation of the Philippines, the image is but one of thousands that contributed to a visual archive of the empire's new Pacific possession. Scientific personnel embarked on large-scale visual documentation of the archipelago that included Philippine peoples, landscapes, and flora. Images from expeditions circulated through public lectures or by way of government documents, serials, and popular monographs. Brief captions written on slides could point to locations or plant names but hardly much on the local people, usually men, hired during an expedition.

Photographs commonly depicted the relative size of unfamiliar flora for colonial and imperial audiences. Personnel and assistants posed alongside towering tree trunks and dense forests to strike intellectual curiosity or to encourage business investment. Through the lens of the sovereign vernacular, the lantern slide conveys not only the politics of imperial photographic subjection but also an expression of plant knowledge. The field assistant is not standing to show relative height. He reclines, at a moment of leisure, looking down at the photographer in the clearing. Beyond respite, such a pose suggests the field assistant's familiarity with the balete, ability to climb it, and acquaintance with its unique structural integrity, enough to rest against its aerial roots. For a moment, the field assistant, a hired collaborator in colonial plant surveying, demonstrates an ease to which the would-be viewer may not have access. At once, the colonial science is made through the technology of the photograph and local labor, and unmade by the field assistant's repose and a knowing uneasily tapped. The balete, which typically begins as an air plant with roots above ground, eventually girdles and suffocates its host tree. In this photo, the balete, which cradles the field assistant, is in the process of its slow, gnarled encircling—a species' encroachment on what had been a capitalist endeavor during the Spanish regime and would continue to be under the United States. It is vernaculars like this that serve as points of departure in order to set the historical stage and the broader conceptual moves ahead.

Setting the Stage

The Philippine Convergence

This book is situated in the last four decades of Spanish colonization (1858–1898) and the first four of US colonial occupation (1898–1935). In 1858, the Spanish colonial government established the Jardín Botánico de Manila. The institution and its Philippine-born and foreign personnel marked a significant shift for Spanish imperial science and for colonial intellectual production on plants. A revision of a flora on Indochina, a project begun by a leading US colonial botanist during his two-decade station in the Philippines, was published in 1935. The work features key nomenclatural considerations that punctuated ongoing international botany debates and reflects an enhanced effort on the part of the US to explore neighboring colonial terrain. Both moments and their intervening years saw several developments central to the legacy of the science in the archipelago.

The Philippines' historical trajectory and location make it a special locus of analysis for a sustained study of botany. Prior to Spanish contact, the islands were the site of politically independent animist societies and Muslim sultanates linked by trade and war. Maritime polities, such as those located in Sulu, Cebu, Manila, and Tanjay, engaged in terrestrial, interisland, and regional networks that stimulated the circulation of goods from the ports of Majapahit, Champa, Ayutthaya, and Java among others.[3] Oral traditions predominated, and though languages had scripts, no textual material has survived. Iberian fleets, driven to discover competitive trade routes and access to the lucrative plant commodities of the Maluku Islands, arrived at what would become "Las Islas Filipinas," named after Spain's Philip II, in 1521. In 1565, the Spanish established the first permanent settlement on the island of Cebu, administered by the viceroyalty of New Spain. Mass Catholic conversion, enslavement, labor and military conscription, wars, and political subjugation over the next three centuries ensured that the process of colonization was neither peaceful nor categorically unquestioned.

Spanish hold over the colony weakened by the nineteenth century. An influx of capital and business enterprise coursed through the archipelago, especially with the end of Spain's trade monopoly in 1815. Liberal politics ascended within the broader ecosystem of political upheavals in Latin America and in Europe. Creole (referring to a Philippine-born Spaniard, also originally known as Filipino) agitators embraced the ideals of revolution in the face of an ennobled friar class and differential political rights. Reformist and anti-colonial movements that had been developing since the early nineteenth century became more visible by the century's end. A crop of middle- and upper-class Philippine-born mestizo (mixed-race of Chinese or Spanish and native parentage) and native intellectuals produced literature, serials, artwork, and intensive studies of the islands, marking what can be considered the "Filipino Enlightenment."[4] No longer reserved for a European class, the "Filipino" identity became a patriotic one for an entire people.[5] With this self-fashioning, a burst of politically revolutionary and liberal activity critiqued the Spanish administration and advanced a new protonational self-determination. Yet what climaxed in the Philippine Revolution

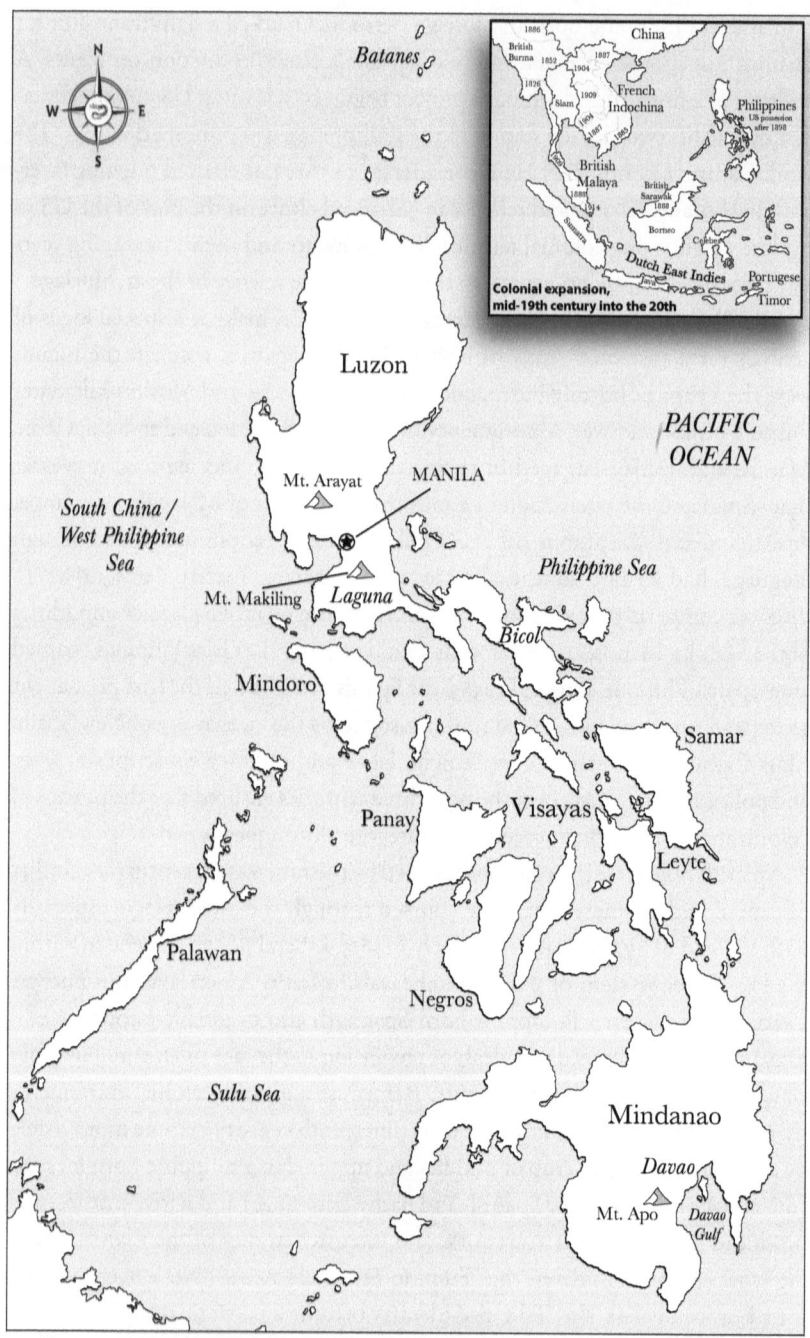

MAP I.I. Map of the contemporary Philippines. Inset map demonstrates European and US colonial expansion from the mid-nineteenth century into the twentieth in the region that would become Southeast Asia. Reference map courtesy of *Philip's Atlas of World History*. Cartographic design courtesy of John Wyatt Greenlee, Surprised Eel Mapping.

of 1896, the first far-reaching, multiethnic anticolonial revolution in Asia, was immediately followed by the arrival of another foreign colonial power.[6] At the conclusion of the Spanish–American War in 1898, the United States purchased the Philippines from Spain through the Treaty of Paris. Formal colonization continued through the mid-twentieth century. After the United States granted commonwealth status to the Philippines in 1935, the colony obtained independence in 1946 at the end of World War II.

To foreign colonials from both Spain and the United States, the Philippine geography was a strategic asset: its position, a prime location for trade and military outposts, and its natural resources, an oft-cited storehouse of wonder and wealth. At the same time, the Philippines' location within the Pacific Ocean made it a rather distant colonial holding from its imperial metropoles, unlike Spain's American colonies or the United States' eventual Caribbean possessions. Fleets of foreign colonials could not land at its several key ports at the same volume as they arrived at La Habana, San Juan, or San Francisco de Quito. The presence of neighboring imperial powers in the Dutch East Indies, British Malaya and Burma, Portuguese Timor, and French Indochina also reduced the likelihood of colonial expansion beyond the archipelago's shorelines. In earlier centuries, nearby empires spelled fierce competition.[7] By the turn of the nineteenth and twentieth centuries, as this book shows, their proximity meant more collaborative potential.

ENTANGLED EMPIRES, EMPLACED SCIENCE. Existing scholarship has painted the two empires as exceptionally different. Indisputably, Spain's first motives for imperial exploration in the sixteenth century were distinct from those that prompted US overseas expansion in the late nineteenth. The global milieu within which these empires operated further affected the level of interimperial hostility or cooperation they practiced. Their programs of race, education, and bureaucracy also diverged.[8] For the Philippines, scholars have also distinguished Spanish colonial science from US. For the anthropological sciences, for example, the Spanish prosecuted less ethnological research compared to other European empires of the late nineteenth century, including the nascent German Empire.[9] The US Empire, on the other hand, ballooned with anthropological investigations that have been the basis for prolific scholarship on the early twentieth-century colonial Philippines.[10] However, a particular brand of simplification has eroded the complexities across the two empires' scientific traditions, often to the detriment of Spanish-era scientific discourse, in service to US triumphalism, and at the expense of local intellectual production.

When the United States assumed control of the archipelago, colonial botanists joined in the chorus of critics to lambaste "Spain's decadence" and its failure "to grow intellectually." Consequently, the Philippines, until the dawn of US Empire, only made scientific contributions "through the mouths of Spaniards."[11] The US colonial tendency to represent the Spanish period as bereft of science left a stubborn imprint, made manifest in Philippine historiography that can oversimplify the colonial past in general and colonial science in particular.[12] Iberian science has faced the same slanted treatment elsewhere.[13] Recent scholarship of the Philippines has disqualified these early accounts, tracing carefully how the purported success of US colonial science rested on the original infrastructures laid by the Spanish and by the local intellectuals who trained in the nineteenth century.[14] With others, therefore, I continue to complicate this distorted interpretation.[15] A focus on the Philippines in the decades surrounding 1898 offers a more symmetrical analysis of two regimes. In this book, I highlight both regimes' approaches to botanical science, their turn to more region-wide interimperial collaboration and research, and more significantly, the contributions of local actors to such work. I probe how a place like the Philippines constituted the "very conditions of existence and enunciation of the [colonial] knowing subject."[16] Such an attempt reincorporates and re-localizes important knowledge-claims that can too easily be divorced from *place*. Emplacement, writes Harri Englund, sees a subject as "inextricably *situated* in a historically and existentially specific condition." Quoting the work of Edward S. Casey, Englund reminds, "'We are never anywhere, anywhen, but in place.'"[17] This approach unseats science, and especially imperial sciences, as ambiguously globalist forces absent of particularity.

This book, therefore, is about more than interimperial squabbling. Such squabbling has hidden the important social and political contours of the Philippines from the mid-nineteenth into the early twentieth century. Even though colonial agents ostensibly brought botany to the Philippines, local conditions and actors (human and vegetal) shaped the course of the science. I cast light on the Philippine-born field assistants, craftspeople, illustrators, and intellectuals, many of whom lived and worked across the historiographical divide. They transformed plant fibers into woven thread, sketched sepals, and warned of spirits in the forest. Their knowledge and labor contributed to the colonial scientific outfits. Their fluency with plant life exhibited knowledges that predated, existed beside, and transformed with the colonial encounter. Particular plant varietals and species, some understood to be endemic to the archipelago, propelled research and commerce for some foreign botanists. For some local actors, they constituted lifeways, political aspirations, and independently developed knowledge of the surrounding environment. Furthermore, as excellent scholarship has drawn attention to the role of the arts and what may be considered the humanities and social sciences in nineteenth-century Philippine nationalist thought, I maintain that botany had a place in such a profound political current, too. This history may resonate with other accounts of colonial botany elsewhere. At the same time, I uphold a politics of difference that recognizes the heterogeneity of historical experiences that, while perhaps generalizable to the larger colonial world, surface from the specificities of place, time, and source material.

Knowing Plants

Exploring botany requires understanding the science historically as a constituent part of a cluster of knowledge systems and not simply a "metonymy of knowledge" itself.[18] Science, or the amalgam of various disciplines that seemingly emerged in Europe, could only attain such a premier position because of its co-constitution with empire and colonialism.[19] I deploy the word "science" to summon the Anglo-European history of which the discipline of botany is a part. The philosophical fervor with which botanists strove to order plant life erected a way of knowing that could hierarchize all other ways. European imperial powers' expansion into Asia and other parts of the world in the early modern period engendered a belief that the world's plants could be cataloged to establish a universal register. The expectation that Carl Linnaeus (1707–1778), considered the father of modern botany, and his apostles could achieve this aim ran headlong into a diversity of ways people understood—and in several respects, continue to understand—plants in Asia and, as this book will cover, in the islands of the Philippines.

I use "ways of knowing" interchangeably with epistemologies and knowledge systems. The phrase emerged in early feminist theory, and I apply it to recognize the context-dependent, experiential processes through which knowledge is made.[20] In this application, I see Anglo-European science with its attendant disciplines as ways of knowing, undivorceable from the contingencies of history and subject-position. The numerous insights into plants that I also cover in this book fall under this larger umbrella category. I hesitate to call or categorize such insights as "sciences." While excellent scholars in feminist science and technology studies and Indigenous studies have argued the place of Indigenous knowledges as sciences in their own justifiable right, particularly with present-day case studies, I consider how the historical source material of my project sharply distinguished between who was a scientist and who wasn't, who practiced science and who didn't.[21] These divisions were stark but at times slippery. It is in some of these slippages that I am able to pinpoint botany's making and umaking. My redefining of the vernacular, as I discuss later, allows what I believe to be a more potent historical intervention into the acute intellectual hierarchies once defined by practicing colonial botanists.

Considering botany as but one knowledge system also emphasizes the rather localized dimensions of its historical development in key sites in Europe and in colonial contexts such as the Philippines. This framing insists that all knowledge systems have identifiable localness: the political contexts, social contours, and material structures that may impact knowledge production. Moreover, this approach underscores the variety of knowledges of which plants were a part and how such systems interfaced, combined, or stood in contention. As Helen Watson-Verran and David Turnbull have noted, the epistemic authority and credibility of Indigenous peoples might be readily labeled as "closed, pragmatic, utilitarian, value laden, indexical, [and] context dependent," as opposed to the virtue of scientific objectivity.[22] In this book, even if botanists working under the banner of international botany aspired for a certain brand of objectivity, their efforts were no less pragmatic, value laden, and context dependent.

WORLDS OF PLANTS. In artifactual, visual, written, and oral records, plants have long been instructive and significant to human knowledge systems. Forbidden fruit, blessed grass, and tutelary trees comprise stories of earthly origins. Extraordinary plants, for example, populate the universe of Mount Meru, the hallowed mountain of Hindu, Jain, and Buddhist cosmology, at times acting as icons and reminders of a non-anthropocentric moral order.[23] Reflecting this sentiment, the Hindu *Mahābhārata* epic, compiled between 400 BCE and 400 CE, contemplates trees as pleasure- and pain-feeling sentients.[24] Plants have also operated as channels between human and spiritual realms, such as the lotus medallions engraved on temple ceilings of Angkor Wat that mark both cosmopolitan power and sacred space.[25] Likewise, the monumental banyan tree has been revered as the location where Siddhartha Gautama gained enlightenment, and where, in animist belief, guardian deities and spirits live.[26]

Alongside their cosmological and archaic importance, plants have been prominent for nomadic communities, agriculturists, forest gardeners, and medical practitioners. Archaeological research suggests that early inhabitants of Tabon Cave on the island of Palawan may have relied on plant technologies, such as pliant fibers for basketry, weaving, and boating, as early as the late Pleistocene.[27] In the textual record, plants' healing virtues have comprised a significant part of intellectual inquiry, trade, and practice. Classical-era documentation reveals how Indic, Chinese, and early Hellenic medical traditions relied considerably on plants for elemental or humoral wellness. The early Tang Dynasty sponsored the compiled pharmacopoeia, the *Shen Nong Ben Cao Jing*, and a systemization of *materia medica* knowledge became a central element of the bureaucracy's regulation of the practice of medicine.[28] Scholars in antique Mediterranean Europe participated in analogous developments.[29] Thinkers of the Persianate Age, like Abū al-'Abbās al Ishbīlī (d. 1239), studied plants' medicinal properties together with their morphological features in an empirical manner.[30] Over centuries, trade and tributary relationships facilitated the movement of medicinal plants like frankincense from the Arab world through Srivijaya to China, or conversely, a Chinese varietal of cinnamon to Arab apothecaries, and their enfolding into local medical custom.[31] At the same time, these exchanges included firm recognition of difference, as in the case of Vietnamese practitioners' resistance to aspects of Chinese medical practice to protect a distinct "southern" tradition.[32]

Land and maritime networks accelerated the exchange of plant material and the knowledge of their commercial potential. The trade in natural objects—vegetal, animal, and mineral—wrought prestige: aromatic resins, pangolin skins, and unblemished lapis lazuli traveled from ports to the palms of royalty. Asian and European courts salivated for exotic plants and the cachet they brought as worldly, collectible objects. These expressions of aesthetic and intellectual dominance increased certain plants' economic value, especially those traded from present-day Southeast Asia.[33] Clove from the Maluku Islands, for example, not only caught the attention of early modern Iberian and Dutch traders. Third-century BCE Chinese courtiers had been expected to have the breath-freshening dried flower buds in their mouths when in the emperor's presence.[34] Plant products for ritual ceremonies, feasts, adornment, and visual indulgence were brought to Buddhist temples, Javanese celebrations, Arab perfumery workshops, and Habsburg curiosity cabinets.

A systematic ordering of vegetal life arose with courtly expansion and increased foraging. Regal imperial posturing and resource extraction predicated stricter classificatory systems, particularly as naturalists encountered unrecognizable flora. As empires obtained neighboring or distant lands, ordering plant life could wield material, moral, and aesthetic power. Put another way, the taming of nature (and the nature of others) could evidence magisterial control. The Mughal emperors of South Asia, for example, contracted the production of intellectual knowledge on plants and gardens to govern the natural environment. Visualizing and establishing royal gardens was an imperial approach that simultaneously reordered the environment and mapped authority onto it.[35] The accumulation of plant knowledge and plant material surged with the intensification of European empires from the fifteenth century onward. "Exotics" from Asia, the Americas, and the European "margins," such as Scandinavia and northern Muscovy, prompted Renaissance naturalists to describe with urgency.[36] Naturalists joined maritime explorations to record the most bizarre, the most useful, and the most marketable. Such was the case for the earliest Iberian voyages that necessitated novel observational and descriptive methods of peoples and landscapes, which catalyzed an unparalleled European interest in natural history.[37]

LINNAEAN SYSTEMATIZATION. While classical and early modern scholars worldwide studied and wrote on plant life, botany in its Anglo-European formation, with which this book deals, reached a defining moment with the work of Linnaeus. Linnaeus tried to fortify an "autonomous science of plants," which had long been practiced and perceived as an auxiliary of medicine. In 1751, his *Philosophia botanica* rigidly distinguished "real botanists," a class of plant illustrators, collectors, classifiers, and describers, from "botanophiles," or medical practitioners, gardeners, and enthusiasts, who did not advance systematization. He and other botanists, backed by stately royals and trading companies, patronized a vast web of collectors and like-minded practitioners to re-entrench the discipline with new plant material and systematic plant investigations.[38]

Linnaeus, alongside his French contemporaries, sought a standard language, set of empirical rules, and quantifiable approach that coincided with European Enlightenment-era tenets.[39] Plant material from surveyed, conscripted, and colonial territories challenged the philosophy of the discipline. At the same time, the deluge of novel material motivated European naturalists to create more localized florae (those on their own backyard, so to speak) in an effort to discover what could be considered "indigenous" to Europe itself. Linnaeus began his career studying his local and national flora, only to eventually dismiss local florae of Europe and elsewhere for their "methodological eclecticism."[40] His *Species plantarum*, published in 1753, built on a genealogy of centuries of botanical writing and popularized a methodological consistency for European botanical studies. Prior to Linnaeus's work, Renaissance naturalists had not necessarily espoused a taxonomic scheme for plant life. As Brian W. Ogilvie has detailed, Linnaeus stood apart from his predecessors because he differentiated "species" (a single taxonomic unit for similar organisms) from "genera" (a class of organisms with similar characteristics of which several different species can be a part) based on floral organs and proposed a system that could anticipate previously unknown plants.[41]

Linnaeus attempted to identify all plants with a Latin-based two-part naming scheme, as others had before. Yet his publication did away with an unregulated polynomial system that left species with a dribbling mouthful of Latin. Communities still rely on this two-part system, which consists of a plant's genus followed by its species designation, like *Ananas comosus* for piña or pineapple. Linnaeus also promoted a system of arrangement following the "sexual" characteristics of a plant: "female" and "male" organs and anthropomorphized nuptiality practices (some unbelievable to the time) could reveal similarities across species.[42] Contemporary botany still uses these Linnaean designations, despite the well-known limitations of his "artificial" approach—that is, the prioritization of specific plant morphological structures to indicate relation.

European botany and the social position of "the botanist" flourished following *Species plantarum*. Lexical research reveals that the word "botany" was more than fifty times more frequent than the word "botanist" in the first half of the eighteenth century. By the early nineteenth, botany was only three times more frequent. Interpreting these findings, René Sigrist suggests a growing "affirmation" of the botanist. Linnaeus's systematics grew in popularity as botanophilia among the broader public became more widespread. Alongside these trends, public and private concern with botany's value increased as observers identified profitable connections between economy and the science. As much as there existed a refined "Botanical Republic"—an "ego-network" of correspondents with Linnaeus at the center—so too was there a brute commercial lust for plants.[43]

The discipline advanced as European states realized material gain from domestic agricultural development and imperial expeditions. Botany assisted in the social governance of territories, cultivating a productive, patriotic class of human subjects, domestically in Europe and abroad.[44] It also offered a methodical approach to agriculture. Research on economic plants targeted species that could be introduced and acclimatized to new landscapes, selectively bred for their most marketable qualities, and grown in mass quantities. Cinchona, nutmeg, cinnamon, pepper, tobacco, and cacao, to name a few, tantalized merchants and botanical investigators. The Spanish, Portuguese, Dutch, English, and French empires guarded their colonies' natural resources under the threat of what Londa Schiebinger calls "biopiracy" from rival empires.[45] Competition over trade networks to acquire these species—and the lands upon which they grew—escalated.[46]

A PROFESSIONAL, INTERNATIONAL SCIENCE. Anglo-European botany confronted epistemic distress in the nineteenth century. The philosophical and practical concerns of the previous century had been worsening. Continued colonial exploration inundated botanists with nontemperate plant material, which did not neatly conform to temperate-plant-driven systematics. Botanists tried to fit new tropical material, for instance, into genera with which they had more familiarity, incapable of rectifying plant diversity within their arrangement. The number of botanizing practitioners also increased, as natural history enthusiasts, travelers, Linnaean protégés, and artisans from varied class backgrounds took to the field.[47] At the same time, "botany's ancien régime" could not readily discipline the influx of plants or new practitioners.[48] Publication venues proliferated. New Latin binomials went unaccounted. "Amateurs" set foot, some driven more by sacrament and self-improvement than by systematics.[49]

By the mid-nineteenth century, the increase of botanical practitioners pushed self-defined elites toward stricter delineation of what botany was and who practiced it. Furthermore, a continental European approach to systematics had slowly begun to eclipse Linnaeus's northern European artificial system. This approach, championed by Swiss botanist Augustin Pyramus de Candolle (1778–1841), considered all morphological structures of a plant rather than one isolated characteristic. A new botany thus emerged.

Victorian Britain viewed the ancien régime as polite, unserious, and ladylike. Prevailing gender ideology from the previous century branded the discipline as fit for feminine urbanity in middle- and upper-class society.[50] Early commentators cautioned against the unlearned and the "fair sex," who stood to defile botany and its pure language.[51] By the middle of the nineteenth century, "professionalization of botany," argues Ann B. Shteir, "meant its masculinization as well." The formally trained, male botanist ascended as the prototypical professional. Botanical writing, too, shifted direction. Discourse divested of individuality, personal style, and traces of the feminized "familiar" more widely became the standard.[52]

The establishment of the International Botanical Congress (IBC) formalized the discipline even further; hence the organization's role as the backdrop of this book. The IBC's development over the course of several decades modified the self-declared ethos of imperial botany. This modification occurred within the milieu of the "consolidating imperial world" that saw the circulation of politics, ideas, and scientific knowledge beyond the axes once limited to metropoles and their colonies.[53] Spanish and US colonial botanists positioned themselves in dialogue with other colonial and metropolitan scientists as intellectual collaboration became self-consciously international in scope. Convenors of the IBC set out to standardize nomenclatural rules, herbarium norms, botany instruction, a master bibliography, and subfields within the discipline.[54] The IBC catered foremost in its original conception to the old and emerging empires of Europe and came to include the United States more centrally as the country ascended as an overseas imperial force. Under the rubric of "Olympic internationalism," botanists championed the nations they represented in support of an international fraternity first largely limited to the Anglo-European world.[55]

The Moves of This Book

Redefining the Vernacular

Redefining botany's vernacular requires knowing the origin of the term in both its general and intellectual usage. The word's etymology and its study imply uneven relation. The term, which has a possible proto-Indo-European origin, comes from the Latin *verna* for "a slave born in the home of his master." *Verna* became the Latin *vernaculus* for "domestic" or "native" before emerging in English in the early seventeenth century as "vernacular," or that which uses the native language of a country or district.[56] The term carries a profound association with servitude and steep inequality. Based on its provenance, the vernacular is relational, necessitating someone or something enslaving and superior. That the domestic or native had been constituted under the ownership of a master reveals the material and the social histories undergirding the vernacular's formation. The term, to put it another way, has always reflected scales of power, emplacement, and indigeneity.

Philologists, linguists, and classicists have typically contrasted the vernacular (both in its nominal and adjectival forms) with the languages of power and high intellectualism, yet vernacular production—in its oral, textual, and cultural forms—historically maintained its own spheres of influence, which captivated intellectual elites. Humanists of Renaissance Europe eagerly collected relics of popular culture, heritage, and language. Likewise, elites upheld local traditions as emblems of a venerable past in the face of fast-centralizing states. By the early nineteenth century and with imperial expansion, European intellectuals treated the peasant, the provincial, and the primitive as relics "to be recorded in text but eradicated from practice." In tight correlation with the development of the field of folkloristics, the vernacular was both the crossroads of societies' lowest social strata and the location of revolution and social contestation.[57] Scholars in Germany, the Philippines, Japan, and elsewhere from the nineteenth century through the early twentieth investigated the vernacular as sites of folk life-worlds in tandem with dethroning a particular brand of European elite exceptionality.[58]

Parallel developments in imperial sciences systematically studied "primitive knowledge" in colonial territories. The intensification of the discipline of anthropology into the mid-twentieth century legitimized science practitioners' sifting for practices and epistemologies unlike—and in several ways, perplexingly similar to—their own. Helen Tilley has investigated how turn-of-the-century Anglo-European intellectuals who studied African knowledge systems reinforced their own epistemological superiority by cherry-picking information.

Instead of relying on the modifiers these scientists deployed, such as "primitive," "traditional," "folk," and "ethno-," Tilley suggests that historians of science use "vernacular science" to refer to self-identified scientists' investigations into such other (and Othered) ways of knowing.[59] Juno Salazar Parreñas has argued similarly within the context of Southeast Asia. Instead of applying the term "Indigenous," a word that carries specificity to North American contexts and can potentially convey mere contradistinction to settler-white colonialism, "vernacular ideas" communicates the multiple ways of knowing that have bearing on historiography and ethnography.[60] Tilley's and Parreñas's proposals not only emphasize the contingency of knowledge-claims and practices of knowing but also reflexively acknowledge how investigations continue to be prescribed by scientists and academic elites from North America and Europe.

BOTANY'S VERNACULAR. In botany's long history, Europeans translated Latin to and from linguistic vernaculars. Among Renaissance naturalists, Latin operated as the *lingua franca*. Yet, herbals in Flemish, French, and Italian, among others, ensured a steady translation enterprise for those who could make use of *materia medica* but could otherwise not access Latin.[61] Even if Renaissance naturalists used Latin to ensure a bigger reading audience and to correspond with fellow intellectuals, their works followed what we may consider today a kind of folk taxonomy. These naturalists did not advocate for any single classificatory scheme for plants, but their work fed into a universalizing tendency as methods of observation changed. Herbaria and botanical gardens prompted new observational techniques that required removing plants from their ecological contexts—a key element to folk taxonomists' plant identifications. These techniques boosted the number of species and genera that could be compared among naturalists, paving the way for Linnaeus and others to theorize how to classify such an abundance of plant material.[62]

In spite of botany's emergence from folk taxonomies, the vernacular has long implied linguistic and intellectual hierarchy. *Trivalia* (the commonplace or vulgar) in Linnaeus's *Species plantarum* referred to all non-Latin, "barbaric" plant names.[63] Among Linnaeus's eighteenth-century contemporaries, *vernaculus* described common practices or even that which could be considered endemic.[64] English versions of Linnaeus's work translated species' non-Latin synonyms as "vernacular ones."[65] Botanists used the term fearing pollution of "the Linnaean language" by one's vernacular, which could lower the drawbridge to the "absurdities and barbarisms" at the "choice of the ignorant."[66]

Not all European naturalists and botanists treated nomenclatural vernaculars the same way. Missionizing and secular colonists had varied relationships to local nomenclatural systems and spent considerable effort gaining fluency in local languages. In the eighteenth-century Spanish Empire, creoles dismissed the Linnaean system. José Antonio Alzate y Ramírez (1737–1799), considered a doyen of the Mexican Enlightenment, contradicted laws espoused by European naturalists, finding Linnaean systematics ill-equipped to attend to Mexican plant species, poorly contrived on the basis of resemblance, and morally corrupting for its overemphasis on plants' sexual characteristics.[67] Alzate y Ramírez and naturalist Francisco José de Caldas (1768–1816) in Colombia celebrated Nahua and Quechua taxonomies. In Peru, some creole naturalists advocated for the inclusion of Quechua instruction at institutions that educated students in natural history.[68] The techniques of language acquisition and translation in the Philippines, as I discuss in chapters 1 and 3, served the ends of missionary work and of naturalist investigation, even if later derided by hardline systematists. Furthermore, key figures in colonial Philippine botany, such as the Manila-born Spanish mestizo Regino García y Basa (1840–1916), did not dismiss the nomenclatural vernacular. His racial and class position likely informed some of the ease with which he engaged Latinate systematic botany and local plant names.

By the late nineteenth and early twentieth centuries, botanists still wrote in their own linguistic vernaculars and, as chapters 1 and 6 cover, still struggled to standardize the language of the science. In 1905, the IBC upheld Latin as the premier language of plant naming and description. Conference convenors nonetheless accepted German, English, Spanish, Flemish, Dutch, French, Italian, and Portuguese papers. Wherever possible, presenters needed to summarize their findings in the "international" languages of French, German, English, or Italian. Moreover, every governing motion of the congress had to be in French.[69] Latin proved to be more of an ideal than a shared reality.[70]

These hierarchical conceptions of vernaculars within the science were themselves classed, gendered, and racialized. They mapped onto notions of the ideal scientific practitioner, who in the nineteenth century was envisioned as Anglo-European and male, and onto the litany of laborers and knowledge-bearers assisting his scientific conceits. Like the imagined body of the practicing botanist, one's scientific writing had to be refined and use the tongues of empire. In the Philippines, Latin, Spanish, and English were the principal languages of scientific production under the colonial regimes, and the most prolific authoring botanists were Iberian and US personnel. As more Philippine-born men trained in the science, their authorial practices, social status, and clothing style more visibly distinguished them from those outside of botany's modern pomp, including informants and field assistants hired in situ, which I cover in chapter 5. The vast majority of these men hailed from wealthy families and could write and speak fluently in Spanish (or eventually English), skills enjoyed by a tiny fraction of the population. While Iberian, Philippine-born, and US women contributed to the science as illustrators, plant collectors, and hired and *ad hoc* administrators, they could not enroll in the Spanish or US botany training programs that were exclusively for men.

At the same time, the racial configurations of the colonial Philippines lent a distinctive dimension to scientific production. Racial categorizations were at once politically and ethnologically defined. Unlike in Latin America, a creole *class* never fully materialized since Spanish settlers could not and did not arrive in the Philippines at the same magnitude. Until the mid-nineteenth century, Philippine society distinguished creole status more by way of culture and wealth than by blood quantum alone.[71] By the 1880s and 1890s, the notion of *who* exactly constituted a Filipino began to shift. This porosity of identities contributed to the production of scientific knowledge constructed within the interplay of class and shifting racial identifications. Peninsulares (or peninsula-born Spaniards), creoles, and mestizos dominated scientific writing, and class background secured their access to scientific institutions and training.

Intellectuals of the late nineteenth century both in Europe and in the Philippines, also known as ilustrados (enlightened ones), laid claim to local knowledge. These intellectuals proudly engaged in the most up-to-date scholarship from Europe while brandishing a superiority over, most especially, Spanish intellectual production. Among ilustrados, tracing native knowledge became a major activity, one that showed they could unearth their civilizational past and write expertly on it. They brought into dialogue local knowledge systems and the European disciplines to which their elite stature afforded them access. They deployed the concept of race to establish differences and similarities found within the Philippine population, "albeit unevenly and with exclusions," as Megan C. Thomas caveats.[72] A similar parallel developed in botany. Trained and highly experienced Philippine-born mestizo and native personnel excelled at plant illustration, local nomenclatures, and collecting plants. Compared to their foreign contemporaries, their language skills and familiarity with the environment helped them tap information otherwise difficult to ascertain. These local experts also classed and gendered vernaculars in the colonial Philippine setting. As the following chapters will suggest, these processes were neither straightforward nor purely produced by white colonials.

In light of the term's history, I redefine the vernacular as much more than linguistic expression, as botanists once and may still uphold. Such a limitation avoids the multidimensional human-plant engagements of the past and of today. Instead, I see the vernacular as the insight derived from the varied ways by which people come to know plant life and communicate such knowing. These ways include, and yet still exceed, botanical names, visual artwork, creative writing, intellectual production, material manipulation, bodily knowledge, cosmology, and ritual belief. The following chapters present my expanded definition, taking into account varied experiences with and knowledges made of the plant world: from the culinary textures of rice, through the lyrics crooned to honor a flower, to the haptics of a skirt woven from banana fiber, to name a few. These, I maintain, offer new vantage points from which to examine the history of botany. Given the vernacular's etymology, historical use, and racialized and gendered configurations, the term is generative because of its ongoing recognition of a world unequal. The category of the vernacular could not exist without material inequality and perceptions of difference. At the same time, Anglo-European botany's foundation could not be without the vernacular. Their co-constitution troubles any notion of a "durable" dichotomy and ensures that neither can be fully disambiguated.[73] Disambiguation is itself a practice of putting-in-relation. In these relations, the sovereign arises.

The Sovereign

Even as colonial Spanish and US botanists often wrote in their own linguistic vernaculars, they viewed their work as intellectually distinct from that of nonspecialists (and even one another). What distinguished their vernaculars from the ones I bring to light was their subservience to an Anglo-European systematic orthodoxy. I offer the modifier "sovereign" to emphasize such tension found in the vernacular. In my interpretation, the vernacular does not fully surrender to the sovereign of orthodoxy. At the same time, the sovereign's very existence relies on the vernacular.

A VEXED TERM. Like *vernaculus*, sovereign's etymology emerged from the coconstitution of what was perceived as above and what was below. From the Latin *super*, or above, came "sovereign," a Middle English combination of the Old French soverain and the English reign. In its use, the word has had political, gendered, and territorial valences. As a noun, sovereign refers to rulers, majesties, and in one obsolete form, husbands in relation to wives. As an adjective pertaining to things or qualities, the word describes that which is paramount, principal, or most notable.[74] In one of its most distinguished academic configurations from the twentieth century, sovereign describes an "imagined community" deriving power not from some divine or foreign source but, in fact, from the nation.[75]

A troubled and liberatory essence saturates the modifier sovereign. On the one hand, the term implies enforcement of authority and European styles of governance, classed as elite and gendered as masculine. On the other hand, the term evokes a declaration of power against prevailing hegemony. Indigenous studies scholars, for example, invoke the term in relation to territory, rights, and political theory to counteract historical and ongoing settler-colonial effacement.[76] Simultaneously, sovereign*ty* remains a European term whose adoption and application in post- or decolonial discourse and activism has untold implications.[77] Observing this vexed definition, Joanne Barker traces two declarations of sovereignty: that conceived through the project of white supremacy, and that of Indigenous feminism. For the former, an individual sovereignty takes precedence as state-defined rights protect liberal ideals of property, freedom, and political autonomy. Conversely, an Indigenous feminist declaration objects to such an imperial, neoliberal project that operates in the name of accumulation and is conceived, instead, through ethical concerns of relations.[78] In science studies, the "post-sovereign" has been deployed to signal a more recent moment in which science has failed to approach its universalist aspirations. Scientific experts have lost hold of default supremacy, conceding instead to intellectual conflict.[79] Present-day society exists, one might say, beyond the sovereignty of Enlightenment-era science.

Indeed, the project of an all-sovereign science dates back to the European Enlightenment. That Enlightenment-era practitioners proclaimed a universality of knowledge is now a truism in the history of science. This avowal of universalism took form in the early modern Republic of Letters and continued through the surge of scientific nationalism of the late eighteenth century into the nineteenth, due in large part to the French Revolution (1789–1799) and Napoleonic Wars (1803–1815). Patriotic impulses that fueled nationalist science contributed to the formation of supranational bodies—twenty-three, the IBC among them, founded between 1860 and 1899—that governed the practices, theories, and cooperation of member states. By the 1930s, scientists championed the sovereign of a world-universal science, particularly with the rise of violent German nationalism, a moment I discuss in chapter 6.[80]

For the Philippines, the turn of the nineteenth and twentieth century was a critical colonial, proto-national, and imperial juncture. Politically speaking, the language of sovereignty was on many lips. For polymath writer José Rizal's (1861–1896) *Noli me tangere* (1887), the "sovereigns" of the fictional town of San Diego were neither the most landed nor God himself but rather the feuding, self-serving ensign and friar of the township. Such a satirical stab at the Catholic church and the colonial government implored a different political future for the Philippines. Not long after the *Noli*'s publication, reformist intellectuals and revolutionary insurgents toppled a regime, yet as one imperial power waned, another waxed.

In the following pages, I show botany's importance in this period as it served colonial intellectual enterprise and as it offered a vocabulary by which to conjure important emblems of Philippine knowledge. The environment was a muse, a source of the territorial and intellectual sovereign that I explore in chapter 3. As in the epigraph of this introduction, the stanza from "Ang Mutyâ ng Kasilanganan" paints a setting of floral and faunal opulence. The poet writes of an idyllic "Jewel of the East," echoing the same wonder that befell foreign naturalists upon arrival in the Philippines. But we readers remain distant from that place. We must trust the poet, who writes of "there," contemplating both our location and the poet's. By the time of the poem's publication in a serial openly critical of the US colonial regime, the Philippines had witnessed remarkable unrest. Thus, the stanza not only promises a lush location but also intimates the Philippines' recolonized position in the early twentieth century.

Within this historical context, I uphold Danilyn Rutherford's view of sovereignty as a "value that social actors ... seek to have recognized as their own but never can fully possess"; something "internally disturbed: unsettled, inconsistent, fraught with contradictions, never quite as supreme as it may seem."[81] Whereas Rutherford examines the relationship between "would-be sovereigns" and the audiences they need to secure their authority (a never complete and unruly expectation given audiences' own sovereignties) in Indonesia, I trace a similar dynamic between the sovereign and the vernacular. Like the vernacular, the sovereign's premise is relational: someone or something must be lorded over, or someone or something defies such an assault. In this book, my usage of "sovereign" takes a doubled meaning: the sovereign of imperial botany fed by the vernacular, and the sovereignty of the vernacular. For, as I show, sovereign vernaculars outpaced the aims of Linnaean science, simultaneously making and unmaking the science—advancing its utmost aims or challenging its tenets.[82] Botanists iterated upon the science, formulating and reformulating standards at signs of philosophical and practical distress. Different ways of knowing plants and the actors who espoused them contributed to this need for iteration, pushing botanists to insist what did and did not constitute the science. The title of this book emphasizes this ongoing dialectic. One can only _un_make that which has been made and so on.

SOVEREIGNTY OF THE VERNACULAR. Colonial scientists from both Spain and the United States, driven by international science's obligations, contended with sovereign vernaculars in their local environments. As botanists in the Philippines sought to be interlocutors with the greater scientific world, one gathers a sense that the worlds with which they were contending were numerous. In my reading, vernaculars expressed order, lent meaning to, stabilized, and cohered their participants' lives and experiences. However, my available material, a good deal of which comes from colonial interlocutors, hampers my ability to fully argue the existence of multiple ontologies.

Instead, I embrace an "onto-epistem-ological" approach, as Karen Barad articulates, that sees knowing as a product of the "intra-action" of material (what, as can be gleaned, were understood to be plants by historical actors) and the cultural (the meaning-laden ideas, studies, and projects on such material). Different from "interaction," which presupposes the existence of at least two predetermined entities (for instance, a human-botanist and a plant), intra-action accounts for the emergent character of material becoming. Plants may, therefore, be understood as phenomena "constituted and reconstituted out of historically and culturally specific iterative intra-actions of material-discursive apparatuses."[83] In other words, plants, in my reading, do not exist as independent material entities outside of a botanist's study of them—and vice versa. They made and make each other. Such an approach also enables me to consider how historical actors spoke of or confronted, among other things, the spiritual vitality of a forest or of a textile derived from botanical matter.

For some, plants' physical structures could reveal long evolutionary relationships between species. For others, spirits inhabited living plants and even resided in plant matter as it underwent material transformation. To several observers, foreign and Philippine-born, these latter knowledges were deemed "superstition." In their view, superstition meant the opposite of a modern, scientific intellect. This diametric opposition operated under the presumption that science provided a singular, objective take on reality. In the history of science, but perhaps especially in the history of colonial science, proponents of this singular take confronted other knowledge systems—an axiom in a field largely concerned with epistemic conflict. For the Philippines case, actors cast these systems as quackery or exceedingly irrational, which I explore in chapter 5. My aim in this book is not to pronounce what one critic might allege to be "epistemic charity," an upholding of difference that may consequently deny once colonized peoples the opportunity to change their own knowledge systems.[84] It is, in fact, these sorts of hardened categories ("superstition" versus "science") that make for teleological history, or what Davide Wade Chambers and Richard Gillespie characterize as "pushing back the frontiers of superstition and ignorance, with religion and belief retreating in the face of superior scientific explanation."[85]

Instead, this book examines the interplay of different knowledge systems—how they abutted, how they functioned, and how they perhaps sailed as ships in the night. I look to lexicographic records, drawings, published writings, newspapers, photographs, textiles, and weaving implements beside botany tomes, herbarium specimens, specimen tags, and field correspondence. I necessarily include the objects of other disciplines to paint a fuller picture of engagements with plant life. My sovereign move, therefore, is also methodological. The material analyzed in this book points to the labor and knowledge of informants, artists, commentators, weavers, dyers, field guides, and collectors, whose direct perspectives can appear threadlike across colonial botany documentation. Treading into the disciplines of anthropology, literature, visual studies, and language analysis permits wider understanding of more ways of knowing plants than what botanists themselves acknowledged. At the same time, it underscores the science's discipline-making. The sources I examine do not come without colonial social entanglement, the problems of authorial power, and the recognition that some ways of knowing were actively extinguished or written of disparagingly.[86] As scholars have done tremendous work to correctly stress imperial botany's epistemological, linguistic, and commercial violence, I study botany from a number of angles that neutralize just how all-encompassing the science has been perceived to be.

The Chapters Ahead

I have divided this book into three roughly chronological sections, each with two chapters. Each pair focuses on aspects of contention and scientific escalation, covering Spanish and US presence in the Philippines. All provide accounts of sovereign vernaculars and their distinct role in unmaking and making botany. The IBC figures in the first and third sections to capture the sweep of imperial botany's transformation. The work of Philippine intellectuals and insurgents during the political unrest of the 1890s into the 1900s appears in the second.

Part 1, "A Botany at Its Most Defined," follows the development of Spain's most sophisticated expression of botany in the Philippines in its over three-century hold over the colony. Its first chapter ambles into the fledgling Jardín Botánico de Manila. In its early years, the institution did not live up to its promise as an illustrious pleasure and research garden. Deprived of much metropolitan support, Regino García, the garden's first Philippine-born employee, began systematically arranging the garden's seed bank. The increasing collection included varieties of rice known to grow locally, the near majority of which did not have a Latin name. Despite the varieties' morphological similarity to *Oryza sativa*, the binomial for common rice, those most versed in the grains distinguished them upon sensory and cultural parameters outside of colonial botany's purview. This difference in knowledge systems had been annoying European naturalists for over a century. The problem of the plant variety even factored into the formal standards the IBC first pursued upon its founding. I characterize the interaction between European botanical taxonomy and modes of distinguishing varieties as an "asymptotic taxonomy" to refer to botany's far—but never complete—reach to ascertain the "fluctuating" plant form.

As the Manila botanical garden failed to impress the metropole, Madrid decided to intervene by shifting the oversight of the garden to the Inspección General de Montes or the empire's scientific forestry unit. Under the newly appointed leadership of Sebastián Vidal y Soler (1842–1889), a Catalan forester and botanist, the botanical garden grew into a much larger institution. Vidal's work with the garden and more important, his collaboration with García, who was also a trained classical painter, contributed to the apex of modern Spanish colonial botany in the Philippines. In the second chapter, I cover this history and closely examine their collaboratively produced publication, *Sinopsis de familias y generos de plantas leñosas de Filipinas* (Synopsis of families and genera of Philippine flowering plants), published in 1883. Relying on García's systematic visual ingenuity, which was informed by the local contours of arts education and production, the work became part of Spain's scientific statecraft as it continued to position itself as an intellectually competitive empire in the closing decades of the nineteenth century.

However, as Jorge Cañizares-Esguerra has shown for Latin America, "Once the imperial science of Linnaean botany arrived in the 'tropics,' it took on a life of its own, and it was eventually deployed by local patriot-naturalists to undermine the very goals that Linnaean natural history had set out to accomplish ... namely, to revamp and strengthen the empire."[87] Part 2, "Science in a Place of Flux," therefore, covers a period of extreme political activity: the roar of anticolonial politics and the rise of the US Empire in the Philippines. Chapter 3 situates readers in the political foment of the late nineteenth century, when native intellectuals, workers, and peasants amplified critiques against the Spanish colonial state in efforts toward political self-determination. These years were a time of heightened cultural and intellectual activity among ilustrados. Their writings and creative works drew up gendered, everyday renderings of the sampaguita, which is currently the national flower of the Philippines. The chapter demonstrates, among other things, how Manila-based intellectuals used botany's vocabulary but cast aside its other specifying elements to position the sampaguita as an emblem of unique cultural bearing.

Chapter 4 pivots to the beginning of the US colonial period. After a short-lived glimmer of independence from Spain, the US colonial administration established itself in the colony, erecting institutions of scientific research to, among other objectives, survey the commercial profitability of the islands. US botanists most certainly had a hand in these pursuits. Their writings specifically on Philippine weaving and dyeing demonstrate the type of systematic work of identification they invested in to serve commercial interests seeking to scale up plantation production. In this chapter, however, I also offer a contrapuntal story within this larger narrative of settler-colonial enterprise. I provide an example of a US anthropologist conducting fieldwork among a Bagobo community in the Davao Gulf in Mindanao and the knowledge of weavers this anthropologist obtained. This ambivalent case study—and the attendant objects and information that come with it—pushes against the worldview with which botanists entered the Philippines at the time.

Finally, part 3, "Assembling a Wider Expanse," follows US botany's attempts to master not only the Philippine natural landscape but neighboring colonial ones as well. As US botany expanded in the Philippines during the Philippine–American War (1899–1902), botanists immediately realized the need to rely on Philippine-born field guides, translators, and laborers to fully assess the colony's floral landscape. This reliance on native personnel, however, proved tenuous and, at times, dangerous. Chapter 5 examines the tricky dynamics between botanists and native personnel and homes in on the matter of "superstition" tied to forests and lands among native field labor. US personnel observed the frequency and diversity of superstition, which had impeded proper excavating of Philippine domains. Complaints of superstition, I trace, were not new to the US colonial period or to foreign observers alone. At the same time, I argue that critiques of it were seated within US botanists' own fears of their vulnerability in prosecuting botanical work in places altogether new to them. Philippine-born personnel themselves had their own views on the difficulty of field labor that complicate botanists' early appraisal that most laborers were saddled by "superstition."

Chapter 6 returns to the matter of names but examines how non-Latin plant names served as a methodological ingredient to making sense of Latin nomenclature. I recount the collaboration between Mary Strong Clemens (1873–1968), a US plant collector, and Elmer D. Merrill (1876–1956), one of the most revered US botanists in the Philippines, toward the revision of Portuguese botanizing friar João de Loureiro's (1717–1791) *Flora cochinchinensis* (1790). Merrill relied on Clemens to extract material and local knowledge for the grand revision of *Flora cochinchinensis*, an extensive flora of present-day Vietnam and southern China. A Linnaean taxonomist and presiding member of the

IBC, Merrill critiqued international botany practice that failed to account for local plant names—a position reinforced by his time in the Philippines. For generations, the nomenclatural vernacular necessitated the creation of a global language to bring comprehensibility to the "Babel" of local names, a characterization used by scientists and historians alike. Instead, this chapter focuses on a moment when a vernacular exposed the Latin *babble*: the diachronic capricious use of Latin binomials.

PART I

A Botany at Its Most Defined

I

An Asymptotic Taxonomy

> It would be very annoying work to refer to
> all the species or varieties of rice, both wet and rainfed,
> for there are so many.
> —FRANCISCO MANUEL BLANCO, *Flora de Filipinas*, 1837

Pilit, diket, tapol, pirurutong, dinagupan, and malagkit refer to a distinct variety of rice recognized by different sensory qualities of their grains—at harvest, before cooking, after preparation, and during consumption. Its color, ranging from shades of pearly white to hues of violet, shifts after boiling, and even then, depends on when panicles were first harvested. A fragrance emerges when soaking grains in lukewarm water, while simmering and straining produce a unique scent. After cooking, the variety may have a sweetness on the tongue and a considerable adhesiveness that can leave grains sticking to one's fingers or the back of a wooden ladle. Its mouthfeel can be firm, requiring extra chewing, unlike other varieties with a little more give. Depending on location, time of year, moment in the day, and perhaps even passing preference, the grain can transform into a delicacy of which there are too, too many to list.

Talk of plant varieties is liable to elicit reaction from botanists, whether enthusiasm, exasperation, or bedlam. In botany, a variety is a taxonomic rank below a subspecies, which is itself a rank below a species, defined classically as a group of organisms that can interbreed. Understood to have developed mostly from human intervention, a variety demonstrates only minor differences from its species parent. Subspecies, on the other hand, develop from longer stretches of isolation and environmental adaptation, demonstrating more major qualitative differences. Both are distinct enough to warrant notice and naming, yet similar enough not to be considered new species altogether. Thus, the variety of malagkit is taxonomically written as *Oryza sativa* var. *glutinosa*. The genus and species name, *Oryza sativa*, is marked by its variety (var.), *glutinosa*, characterized by its thick opaque seeds. For the botanizing missionary Francisco Manuel Blanco (1778–1845), recording all the varieties of rice in the Philippines in the mid-nineteenth century was toilsome. By the century's end, at least 144 were known to occur across the archipelago. Today, 120,000 are estimated to exist worldwide. Blanco would be beside himself.

The Varietal Challenge to Colonial Botany

The rice plant had an extensive history across the islands before nineteenth-century investigations. Trade networks and a global cooling event in the third millennium BCE facilitated the southward movement of rice agriculture and farming communities from the present-day Yangtze Valley to Southeast Asia. A tropical subspecies of rice, *Oryza sativa* ssp. *japonica*, likely arrived in island Southeast Asia in the sixth century BCE. Once in the island region, the species quickly diversified, yielding its current genetic range.[1] Rice was a preferred food across the communities that would come to comprise the Philippines. Yet, its consumption was reserved primarily for elite classes and feasts since supply was inconsistent and cultivation was limited.[2] However, the earliest foreign chroniclers noted the immense vocabulary associated with the different stages of production and cultivation, preparation, and ritual practices, suggesting "an indigenous, detailed, fully articulated world of great importance" associated with rice.[3]

It is generally understood that a robust vocabulary conveys specialization: butchers may more precisely name cuts of meat just as musicians may describe their art with more lexical acuity. The same principle applies to people who have long interacted with culturally significant plant life. Folk taxonomies of culturally important species are assorted, fluid, and exceptionally vast—drawn from locally encoded cultural knowledge, values, spiritual beliefs, and aesthetic tastes.[4] As contingent, intra-active processes, sensing and the sensory that may inform such taxonomies also reflect bodily experience, ideologies, and meanings. That two plants that may appear alike to one observer but may altogether be two different entities to another speaks to these processes. Research on species variety has shown that chemosensory cues such as taste or smell, alongside knowledge of growing patterns and environmental conditions, help seasoned farmers distinguish more varieties than names alone. These very metrics for identification shift across generations within the same farming communities. As commodity chains lengthen, however, this knowledge condenses in service to commercial efficiency.[5]

The advent of Spanish colonization transformed Philippine rice agriculture into a more widespread practice and rice consumption into a more frequent phenomenon. The cereal grain supported colonial administrative structures and the clergy.[6] Novel plow technology, heavily cultivated friar estates, and amplified commercialization, especially geared toward export-oriented agriculture, turned rice into a commoditized staple.[7]

Rice also captivated missionary botanizers and secular botanical systematists in the Philippines. Its cultural, dietary, and commercial significance made it an inescapably important species to study. By the mid-nineteenth century, rice became central to the Spanish scientific outfit, which began to pivot from purely agricultural endeavors to plant systematics. This pivot appeared among the first institutional architects of the Jardín Botánico de Manila (JBM), whose displeasure with the Philippines' agricultural output partially motivated a more strategic turn to botany. Botany also became one way the Spanish peninsula asserted its imperial dominance during a century otherwise marked by political upheaval, colonial contraction, and eviscerated coffers. Botany personnel in the Philippines presented systematics as a rigorous, methodical approach to demonstrate the scientific progress. Philippine rice varieties had a place in this vision of progress. As the institutional infrastructure for Spanish science increased in the archipelago, so did European botanists' participation in the science's new imperial order through the formation of the International Botanical Congress (IBC).

This chapter steps gingerly onto the grounds of the JBM to uncover the institutional and intellectual mechanisms for cataloging Philippine rice varieties. To begin, I provide a brief history of the earliest years of the JBM and the political, intellectual, and commercial environments that fostered—and delayed—its start. From there, I detail the collaborative work of Zoilo Espejo

FIGURE 1.1. "Juan Serapio T. Nepomuceno: Tagalog couple pounding rice," ca. 1840. In the illustration, a couple takes turns dropping a heavy pestle into a mortar containing paddy. The mortar could hold dried rice stalks, and the pestle's rhythmic pounding could thresh rice grain, separating the grain from its tough protective hull. Then the contents might be transferred to a flat, woven basket to sift the grain from the husks. Repeating the process could completely reveal loose individual grains. The mortar could also contain unhusked grain, which could be pulverized for various uses, especially culinary. Reproduction permission courtesy of the Filipinas Heritage Library, Ayala Museum.

(1838–1904), the JBM's first peninsula-appointed director, and Regino García y Basa, a Manila-born artist who would take up the longest recorded employment under the Spanish scientific outfit. With Espejo, García compiled the JBM's seed catalogs. These catalogs featured a growing list of locally understood rice varieties, likely assembled by García, that captures an important moment in late Spanish colonial science as botany became more institutionalized in the Philippines and internationally. To the botanical systematist, the varieties appeared morphologically similar to *Oryza sativa*. To communities intimately familiar with the grains, they grew differently, felt different in one's mouth, and had different culinary uses. The catalogs, therefore, presented a conundrum, one that had baffled naturalists for over a century: how could imperial systematists contend with the profundity of varietals—and their attendant taxonomies—that outpaced their own classificatory scheme?

The finished catalogs offer only a trail of grains toward domains of thinking that could have been agricultural, culinary, commercial, and ritual—or any permutation of these and more, but the path is instructive. It reveals, as I subsequently show, how European naturalists themselves historically puzzled through the problem of the variety, perhaps only to admit that the metrics by which they distinguished plant forms were, in some respects, arbitrary. Moreover, the catalogs point to ways of knowing that remained out of reach to mid-nineteenth-century European taxonomists and yet were competently mediated by the locally born García. The sovereign vernacular of local varieties included knowledges that existed beyond foreign botanists' immediate purview. Such varieties, nonetheless, motivated trenchant ordering. In this chapter, I characterize Anglo-European taxonomists' envisioned ideal of ordering all known varieties as an "asymptotic taxonomy," which refers to colonial botany's loftily far—but never complete—hold of varied ways of distinguishing plants. In the parlance of mathematics, an asymptote is a line that a curve approaches but never, *ad infinitum*, meets. I conclude by arguing that as the IBC activated the international machinery for codifying botany standards in the latter half of the nineteenth century, the science had to reconcile its own philosophical limits with respect to plant varieties.

An Intellectual Pivot to Botany

Fissures beset the JBM for much of the institution's early history. These began with the foundational mission of the garden and whether agriculture would be the utmost priority for its personnel and day-to-day function. Over what would become a four-decade operation, the garden pivoted toward botany.

Metropolitan Spain had relied historically on natural history investigation to ascertain the profitability of its colonial landscapes. By the mid-nineteenth century, botany in particular could fashion the waning empire as an intellectually competitive modern state.

Founded in 1858, the JBM originated as a partnership between the administration of Spanish Governor-General Fernándo Norzagaray y Escudero (1808–1860) and the Real Sociedad Económica de Amigos del País (RSE). Whereas funding setbacks and the short careers of Spanish colonial governors-general in the Philippines had thwarted prior plans to establish a botanical garden, a more explicitly imbricated collaboration between the colonial government and a private body ensured mixed funding sources to see the garden's start.[8] The RSE, instituted in 1780 after many suspensions and reestablishments, represented colonial Philippine commercial and intellectual interests and participated in a network of other chapters on the peninsula and in Spain's remaining colonies. Leading up to 1858, RSE members crafted a proposal for a botanical garden to serve as a hub that could jump-start Philippine agriculture.

At the time, privately held agriculture outshone the Spanish colonial government's attempts to tap into what had become a booming export-oriented economy. The financial boons of Anglo–Chinese capital and the inquilino class, or those who were leaseholders of agricultural land, most especially threw the state's agricultural failures into relief.[9] For that reason, the JBM's adjoining Escuela de Agricultura was founded in 1861 to train students—many from Manila as well as scholarship recipients, or pensionados, from distant provinces—in the most advanced agricultural methods. However, Norzagaray presented the RSE's plan to the colony's Gobernador Superior Civil in what his original collaborators considered a plagiarizing move. Repackaged as his own, as one of Norzagaray's loudest critics accused, the JBM became a beautification and sanitation project for the Arroceros district, an area adjacent to the walled city of Intramuros. After the peninsula approved the project, the flood-prone location spoiled even the simplest of agricultural pursuits and overturned the RSE's blueprints, which had proposed the elevated Manila suburb of San Juan del Norte as a more reasonable choice.[10]

In Spain, similar fault lines emerged around the hiring of the JBM's first peninsula-appointed director. Norzagaray insisted that the director be an Iberian with professional training: "No one in the country," he assumed, could fulfill the task.[11] In 1863, Spain founded the Ministerio de Ultramar to reinvigorate scholarly study of the colonies.[12] In consultation with the Real Consejo de Instrucción Pública, the council that oversaw public education, the ministry took eight years to pick a candidate—a delay attributable to the meager number of applicants and decision-makers' displeasure with candidates' qualifications. Candidates generally demonstrated competence in practical agriculture and agronomy. Yet in the council's eyes, they failed to display adequate theoretical training for instruction and leading a botanical garden.[13]

Representatives of the ministry appealed to the council to adjust its expectations, which had seemed unrealistic.[14] Nevertheless, theoretical knowledge of botany remained a critical component of the council's standards, such that some of the peninsula's most revered names in the science, Miguel Colmeiro (1816–1901) and Vicente Cutanda (1804–1866), presided over the JBM's civil service exam. Following a multiday exam in 1866, a peninsular tribunal appointed Zoilo Espejo y Culebra, a Córdoba-born agronomist, to the JBM directorship. Espejo arrived in Manila to join Regino García, who had already been appointed by the colonial government to serve as the garden's head horticulturist. With ten student-workers, García and Espejo labored to put the JBM on its feet.

Limited funding and personnel trampled Espejo's grand hopes to see the fledgling plot transform into an agricultural center. Year after year, the JBM failed to live up to its agricultural promise. The hiccups had also been infrastructural. Heavy rains and annual inundation from the nearby Pasig River waterlogged the soil, disrupting efforts to innovate methods for growing tobacco, sugar, and abacá, among other crops, for large-scale production.[15] The JBM's unimpressive state consequently became the butt of saucy humor for decades, appearing perhaps most unsympathetically in José Rizal's *Noli me tangere*, as the protagonist's wistful affections for Manila are dashed by the garden's unsightliness.[16]

After a verbose appeal from Espejo to renovate the garden and its school in 1869, Spain bluntly refused, citing budgetary limitations.[17] To the metropole, more botanical work could prove the JBM's value. The Ministerio de Ultramar suggested that Espejo "make the proper applications of *botany* to agriculture, and plainly but conscientiously explain the part of the science most related with the crops that could be of instructive benefit to the archipelago."[18] Compared to Cuba, according to the ministry, the Philippines had been underperforming in the production of high-quality exportable crops.[19] Iberian and colonial officials increasingly compared the two colonies in the second half of the nineteenth century, amid efforts to reassert power over Spain's last overseas holdings.

This reassertion of power, tied to the metropolitan liberal foment of the early nineteenth century, eventually impacted botany in both Spain and its colonial territories.[20] During the Napoleonic Wars, the armies of Napoleon

Bonaparte (1769–1821) invaded Spain to gain control of the Iberian Peninsula. From 1808 to 1814, Spain and its allies fought against French forces in the Spanish War of Independence. With the country under French occupation, members of the Cortes de Cádiz, Spain's first national assembly, produced the first liberal constitution of 1812, which upheld values of suffrage, government sovereignty, and the right to private property. Yet such efforts to centralize the state, especially after the successful defeat of Napoleon's armies, did not affect Spain's colonies equally.

In the powder keg of patriotic agitation, Spain suffered humiliating losses during the wars of independence in Latin America (1808–1836). With the end of the old political order and the emergence of the Spanish liberal state, the country spent the rest of the century intensifying its colonial relationship with its remaining overseas territories—Cuba, Puerto Rico, and the Philippines— through political, economic, and social means.[21] Furthermore, the closing decades of the nineteenth century witnessed a "second wave" of emerging overseas imperial powers like Germany and Italy alongside the older empires of France, Britain, the Netherlands, and Spain.[22] Spain was not in a position to expand its imperial geography, having been "largely in a defensive position" with its colonies due to the country's poor financial standing, administrative capacity, military ranks, and international alliances. It did, however, invest in intellectual innovation alongside its imperial peers that would come to characterize this "second wave."[23] The JBM and other botany-focused initiatives in the Philippines were part of this milieu of imperial maneuvering.

A marked preference for botany on the peninsula coincided, as well, with the declining power of Isabella II (1830–1904), who reigned from 1833 through 1868. Though Spain had historically excelled in botany explorations domestically and abroad, crown funding for Spanish botanical institutions was inconsistent during Isabella II's rule.[24] Her overthrow and the Glorious Revolution of 1868, after which "the liberals opened the intellectual floodgates, if only briefly," ushered a new momentum for Iberian botanical institutions and publications.[25] During the Sexenio Democrático (1868–1874), or progressive interregnum before the Bourbon Restoration that saw the royal ascension of Isabella's son, Colmeiro assumed directorship of Madrid's Real Jardín Botánico (RJB) in 1868. He and other botanists, such as Blas Lázaro é Ibiza (1858–1921) and Antoni Cebrià Costa i Cuxart (1817–1886), comprised what has been called the "intermediate phase" to revive the science on the peninsula.[26] With this brief opening of intellectual pursuits, scientific study of the Philippines intensified. These political and intellectual shifts impacted those working in Manila as well, motivating increased study of rice varieties, among other botanical topics.

Varieties of *O. sativa*: The Seed Catalogs

Missionary dictionaries reissued in the nineteenth century and *Flora de Filipinas* (originally published in 1837) by the naturalist-friar Blanco provide an assortment of rice varieties that were in use among particular informants in the Philippines. Missionaries had maintained a lengthy tradition of producing local vocabularies and grammars, especially in service to Catholic conversion.[27] In the broader Spanish Empire, local etymologies presumably provided access to indigenous knowledges that missionaries viewed as Adamic in nature.[28] Blanco's *Flora*, the well-studied compendium on Philippine plants, followed this long missionary tradition of natural history writing.[29] With the dictionaries, his work reveals just a sliver of the grain's profuse taxonomies in local parlance.

In Batangas, for example, a reportedly common variety among Tagalog-speakers was bontot cabayo (horse's tail), a possible way to refer to poor quality rice or to hatch the image of an elongated rice panicle. Ilocanos allegedly knew the variety, too, but disliked its coarse mouthfeel.[30] Gamai, for some on the island of Panay, referred to a type of rice, yet the word in its adjectival form also described especially thin or small things.[31] Bolohan, a Tagalog word that translated to fuzzy, white or gray hair, or something of a velvet texture, was not the most prized variety but was regularly planted because of its resistance to blight.[32] Conversely, macasal was a highly valued grain; its root word, casal, denoting the rite of marriage.[33] Quinanda was similarly superior and planted abundantly in southern Luzon on account of its taste. Compared to others, the variety fluffed generously after cooking though required more water than other varieties.[34] Cabog was both a type of rice but also the word for a "hollow-sounding blow," possibly in reference to its aural properties when threshed. Some varieties' names pointed to their own growing patterns: dumali, for hurried or to hurry, sprouted much earlier than others. Some shared names with different types of plants. Tagalog informants used goyoran to refer to a large vine, a type of banana, and a kind of rice. From the root word guyod and its Hispanized verb form "goyor," which in one usage could mean to support or to swarm in unison movement, the term evokes an image of abundant forms of plant life, found as a concentrated whole.[35]

García would later view the period from 1867 to 1873, which corresponded with the peninsula's accelerated work in botany, as a time of heightened study of Philippine rice at the JBM.[36] The grain's domestic commercial production crested during this period, and its eminence as a staple food in Manila was unmistakable. Lowland male labor in rice agriculture had been increasing with the rise of plow technology during the Spanish regime and with the expansion of plots for wet cultivation.[37] Even with the slow rise of rice milling equipment,

FIGURE 1.2. "Cuadro de Costumbres Filipinas. Acto de pilar el palay" (Image of Philippine customs. Hulling unhusked rice), 1877. Reproduction permission courtesy of the Lopez Memorial Museum.

the majority of the population consumed "home-pounded rice."[38] Gendered labor, which had marked rice production in precolonial records, fluctuated by the nineteenth century as the rice plant was intensively commoditized.[39] In the mid-nineteenth century, Blanco remarked that women performed the most demanding manual labor in rice agriculture in certain Tagalog regions while "their cruel husbands" rested comfortably.[40] Costumbrista-style images from the period depicting scenes of everyday life feature women taking up various phases of rice production, ready to wield heavy mallets against unhusked grain (see figure 1.2), winnow chaff, transport earthen jars of seeds, and prepare and vend rice-based delicacies like the Tagalog puto bumbong (see figure 1.3).[41]

FIGURE 1.3. Jose Honorato Lozano, "Puto bumbong vendor," ca. 1852, although the inscription reads "Jose Lozano, ló dibujó. 1850." A puto bumbong vendor stokes the flame beneath an earthen pot while handling a section of bamboo filled—or about to be filled—with glutinous rice. A customer consumes a small plate of the rice cakes while another, with a child, approaches from behind. Reproduction permission courtesy of the Filipinas Heritage Library, Ayala Museum.

Over the course of century, rice consumption rose. In 1857, the colonial government abolished import duties on rice to meet local demand, and importation increased over six-hundred-fold in the span of only two years.[42] Rice became a part of the daily diet of wealthy landholders, merchants, and native nobility. For the working masses in urban centers, rice was as fundamental an accompaniment to meals as root crops and corn had once been, and in some regions continue to be.[43]

As Daniel F. Doeppers has written, "Rice made Manila, and Manila in turn made rice the predominant crop in its more immediate hinterland."[44] Stretches of inquilino-rented lands produced rice for Philippine markets and for export. Domestic rice supply reached Manila through two principal systems: "inner" sourcing through Manila Bay and Laguna de Bay, and "outer" sourcing from production zones in Pangasinan, Zambales, the Ilocos coast, Panay, and Camarines Sur. Merchants and consumers knew of the diverse varieties that traveled along the two systems, as rice reached the capital distinguished by its "microarea" of production. For example, from Pangasinan and the Ilocos region came "bearded" varieties, or awned rice panicles with a fuzzier appearance, which

ASYMPTOTIC TAXONOMY 39

were much less common to the inner zone. The grain's regional identity was maintained in the market, and its varying qualities were differently priced, with Pangasinan and Ilocos varieties yielding a premium. From the outer system alone, Manila saw a deluge of rice. Between 1854 to 1862, as Doeppers relates, the city received enough rice to feed its population for almost a year at a consumption rate of two cavans, or 150 liters, of rice per person.[45]

As the rice market flourished, Espejo continued to tout to Madrid the garden's research and teaching capacity. Rice steadily became a significant feature of such capacity. Espejo remitted a container of seeds to Madrid's RJB, and a duplicate to Paris' Muséum national d'Histoire naturelle, to enhance their Philippine collections.[46] He also published *Cartilla de agricultura filipina* (Primer on Philippine agriculture, 1869), a booklet designed for elementary instruction. The primer, structured in a question-and-answer format with basic agricultural information tailored to the Philippines, posed questions such as: "Are there many kinds of rice?"; "Is dryland rice cultivation important?"; and "Which varieties will be chosen for seed?"[47] Though agriculture yielded little success for the JBM, Espejo and García spent a considerable amount of effort assembling the garden's first catalog of plant material. In 1868, Espejo submitted a list of seeds, in which rice varieties were pronounced, to the peninsula. He boasted it as the first of its kind published in order to facilitate scientific relationships between the JBM and other international gardens.[48]

Indeed, Espejo and García's catalog had an outward-facing appeal. The two eventually established relationships with gardens in Melbourne, Java, Ceylon, and China and in Madrid, Rome, and Paris.[49] No prior secular institution in the Philippines had been in place long enough to openly exchange systematic plant information with as many countries or as far-reaching colonial territories. The manuscripts, folios, and data from prior state-funded naturalist expeditions had seen little public light, a function of state secrecy to protect Spanish imperial and commercial holdings.[50] By the mid-nineteenth century, however, this practice shifted to more explicit participation in circuits of interimperialst intellectual exchange.

Espejo titled their work *Catalogus seminum in Horto Botanico Manilensi* (Seed catalog of the Botanical Garden of Manila), a five-page pamphlet that reflected the JBM's 1867 collection, for which García served as seed conservator. Published by Giraudier in Manila, the catalog, like most botany publications produced in nineteenth-century Spain, followed the Candollian system named after the Swiss botanist Augustin Pyramus de Candolle.[51] The system drew from a French naturalist tradition that Darwinian evolutionary theory and northern European systematics would eclipse by the end of the nineteenth century.⁵² The catalog featured eighty-three families listed with 360 species. In 1868, Espejo and García produced another version with an updated list of 107 families and 497 species, indicating the steady growth of the garden's collection. The 1868 catalog also included thirty-seven rice varieties listed under *Oryza sativa*. By the 1869 catalog, the number of *O. sativa* varieties increased to eighty-one. By 1871, it climbed to ninety-two.⁵³

The influx of rice into Manila in the mid-nineteenth century likely ensured that the JBM could receive a steady flow of regional varieties in lieu of an expansive research garden.⁵⁴ That these varieties came from regions as far north as the Ilocos and as far south as Panay may partially explain the assortment of varietal names from unspecified local languages that appeared in the catalogs. The catalogs did not distinguish between subspecies or varieties, which European botany recognized as different entities by 1870.⁵⁵ Latinized names, such as *pilosa*, *violacea*, *rubra*, and *præcox*, were left uncapitalized, likely following Latin nomenclatural conventions. Non-Latin names were capitalized: those derived from Philippine languages such as binagacay and inuad, acapulco from Nahuatl, and Spanish derivations such as morado, diamante, and señora. García would elsewhere describe some of the names, such as quinanda, as "early varieties" or those that were planted during the first rains in May and harvested in September. Others, like inanod, were rain-fed and harvested in October. Macan was a lowland rice; pinursigui and tinumbaga, upland. Some were glutinous in nature, like binamba and malacquit.⁵⁶

Several of the varietal names likely came from Blanco's *Flora*, which by the 1860s was a well-known reference accessible to Espejo and García. Taste, visual cues, haptic qualities, and agricultural knowledge appeared throughout Blanco's account. He included medicinal applications, remarking, for instance, on malagcquit's effectiveness at softening the hardest of tumors.⁵⁷ Following the genre of the flora, Blanco's publication did not depart much from other colonial and imperial florae, which provided morphological descriptions, measurements, cross-referencing, and informant expertise. In the illustrated reissue of Blanco's work, visual representations of *O. sativa* (see figure 1.4) also resembled conventional botanical images. Its illustrator presented the structures of three rice plants, including their inflorescence, leaf blades, crown roots, and singular grains. Three variety names—binambang, *glutinosa*, and quinanda—appear beneath the illustrations, referring to the three subtly distinct stalks.

Compared to Blanco's *Flora*, the JBM catalogs provide less of a sense of how the varieties differed and who differentiated them. A neatly arranged list substitutes the gustatory and visually descriptive knowledge of the varieties as well as their agricultural specificity. A genre of tabulation, the seed catalog excises

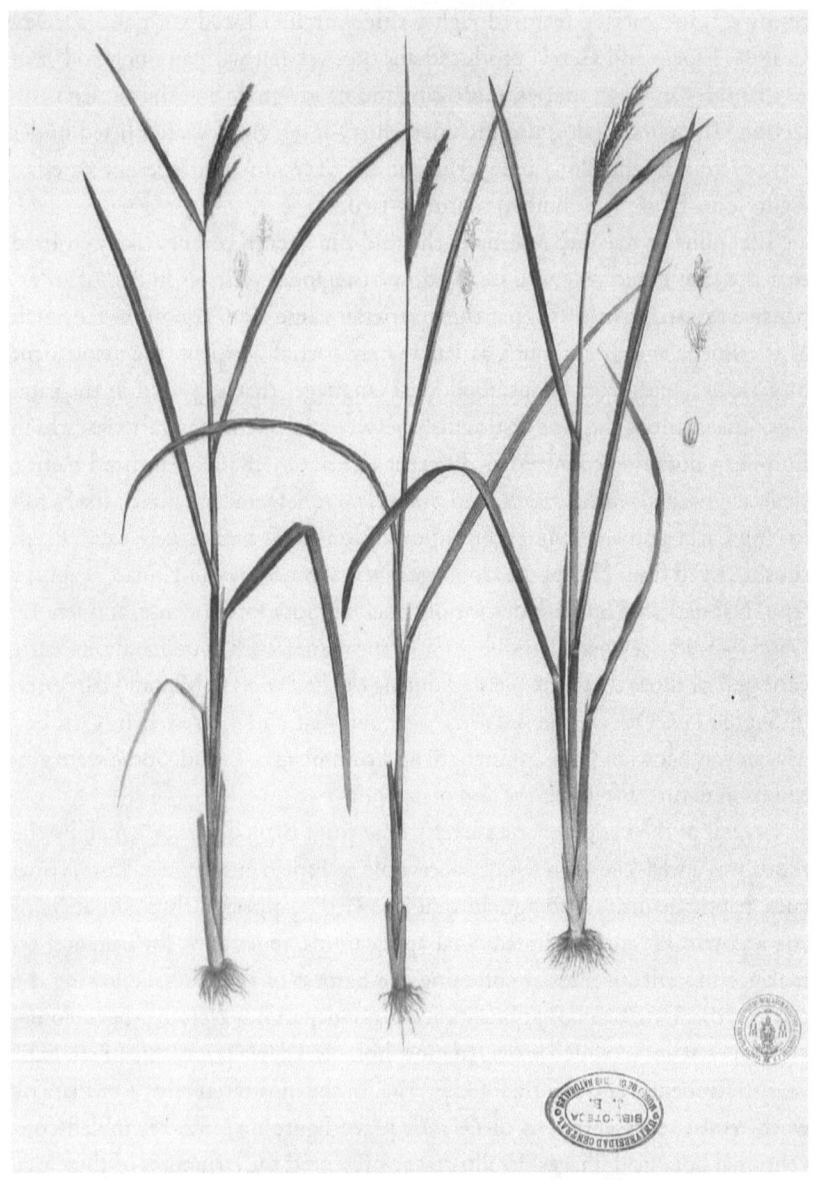

FIGURE 1.4. "*Oryza sativa*—Linn.—Blanco—Miq.," *Flora de Filipinas*, edición 3, atlas I (Barcelona: Verdaguer, 1883). Faint sketches of spikelets sit beside each inflorescence. Rosendo García y Basa, Regino's brother, created the illustration. I discuss the scientific and cultural milieu of illustrations such as these in chapter 2. Image courtesy of Biblioteca Digital Hispánica, Biblioteca Nacional de España.

much contextual information if not for the local variety names that stand prominently amid the Latin. Nevertheless, the local names outweigh *O. sativa*. A reader may be struck by the sheer number, nearly all untranslated (or untranslatable) to a Latinized equivalent. On the pages of the catalog, one therefore finds an interplay of Candollian systematics and the profundity of non-Latin nomenclature—or the sovereign vernacular at work.

García was most likely essential to assembling a vernacular of this kind. Born and raised in the colonial capital to a peninsular father and Spanish-Tagalog mother, García would have been fluent in Tagalog in addition to Spanish. A young mestizo born to a high station, he trained in the capital's finest schools, which could secure his path into government service. Unlike other elite men of his time, García did not live abroad in ilustrado-trodden places such as Belgium, Germany, Spain, and Hong Kong. His artistic talent nonetheless earned him a place to study in Madrid at the Escuela de Bellas Artes de San Fernando. However, political unrest on the peninsula and the death of a fellow pensionado foiled what would have been his departure in 1870.[58] He stayed in Manila, resumed studies in art practice, and continued employment at the JBM to financially provide for his younger siblings, orphaned by the premature death of their parents.

Still, his position as a learned mestizo paralleled that of better-known ilustrados, such as physician Trinidad H. Pardo de Tavera (1857–1925), artists Juan Luna y Novicio (1857–1899) and Félix Resurección Hidalgo y Padilla (1855–1913), or the polymath author, Rizal. Predating some of their works by well over a decade, the JBM catalogs may seem rather pedestrian when compared the political energy of Pardo's, Rizal's, and others' corpora. The catalogs pronounce no overt political critique. Nevertheless, they point to an earlier lineage of Philippine creole intellectual production, able to work proficiently with and between local vernaculars and colonial knowledge production—and in their case, botany's systematics. The catalogs are at once beholden to Candollian conventions yet break from them to show the abundance of rice varietals unknown to outside foreign readers, most especially ones unfamiliar with the sundry local languages represented by the varietal names.

García could have acquired the names from the market, from informants, and from the assistance of worker-students. Espejo's involvement in compiling the varietals is more tenuous. A recent peninsular arrival to the Philippines, he would have lacked linguistic access to the local source languages. His interest in rice might perhaps have come from the grain's commercial and local prominence. A reader of his *Cartilla* could conclude that Espejo may have even preferred wheat as a dominant crop, raising it as the "chief cereal" because of its nutritional content and status as a fundamental ingredient in bread.[59] His opinion, however, had little relevance for the Philippines, a place that had imported wheat flour during the time of the galleons, did not garner a big export market for it in the nineteenth century, and would not allow it to be prominently cultivated on the JBM grounds.[60] The expanding list of rice names suggests García's own growing interest in the topic, which would be corroborated by later sources.

Braided with such practical considerations in Philippine commerce and production, the list of rice varietals provides a sense of the swelling intellectual desire to systematize knowledge of rice. García had not historically been alone in these pursuits. In Spain in the 1830s, the director of the RJB, Mariano La Gasca y Segura (1776–1839), claimed varieties of rice originally sent from the Philippines to be planted in Spain in order to complement a yearslong metropolitan initiative to catalog Spanish cereals.[61] Cereal varieties had become crucial to the work of La Gasca and others associated with *La Ceres española*, an agro-botanical periodical and research program that emerged at the start of the century.[62] Decades after the program's launch, García and Espejo advertised the systematic appeal of the JBM to their colleagues on the peninsula and especially those associated with the RJB. The list implied a commensurability between *O. sativa* and the ninety-two varieties listed alphabetically beneath it. To the metropole, the lists would have highlighted the collecting capacity of the Manila institution, García and Espejo's ability to discriminate rice morphologies, and the precision with which they could distinguish species from varieties. Doing so was much more philosophically challenging for Europeans than it would seem.

Regimenting the "Fluctuating Form"

Through the nineteenth century, European naturalists had no consistent definition for "variety."[63] Even the idea of "species" continued to be a matter of debate.[64] In the history of Anglo-European botany, the variety posed a conundrum to clean systematic arrangement. *Varietās*, the Latin term for diversity or difference, appeared in Carl Linnaeus's *Philosophia botanica*. A variety was a "plant changed by an accidental cause" and "quite often the work of cultivation," whereas "species and genera are regarded as always the work of Nature." As *Philosophia botanica* distinguished those whom Linnaeus considered "real botanists" from botanical enthusiasts, he relegated varieties to the domain of housekeepers, gardeners, and physicians—in his words, the domain of "ordinary life." Humans' intervention in plant development, propagation, and agriculture yielded varieties that he, therefore, excluded from the science. A division of expertise arises in his text such that "no sensible person would readily say that *varieties* are distinct species." Although "lovers of flowers" could carefully inspect plants for "small variations," Linnaeus emphasized that "no sane botanist will enter their camp." Instead, those with a "contagious madness among lovers of flowers" and a "zeal for subtleties" posed a problem so immense that he declared: "The introduction of varieties has done more to contaminate botany than any other thing."[65]

French botanist Michel Adanson (1727–1806), Linnaeus's contemporary and "ardent opponent" in systematics, also addressed the issue.[66] He defined the variety as the difference between individuals of the same species, one that is "accidental and not very durable," as that found among "red, yellow, white, and marbled tulips," for instance. In his *Familles des plantes* (1763), he carefully advised, however, that if what qualifies as a species requires that two individuals must perfectly resemble each other, such a physical resemblance fails to exist: scrupulous comparison yields no perfect likeness. For him, determining the difference between a variety and a species in plants was difficult work in "a field where everyone roams freely."[67] In a separate unpublished note on the idea of the "species," Adanson disclosed that a human observer's sense of resemblance conjured the notion of a species. Varieties, on the other hand, were changes that occurred among species themselves.[68]

Other European naturalists attempted to distinguish varieties from subspecies. At the turn of the eighteenth and nineteenth centuries, they included Jakob Friedrich Ehrhart (1742–1795), Johann Heinrich Friedrich Link (1767–1851), and Christiaan Hendrik Persoon (1761–1836). Achieving standard definitions of the two concepts proved elusive due to shifting usages and language that future botanists would condemn as imprecise.[69] In *Synopsis methodica fungorum* (1801), Persoon, for example, marked varieties and subspecies with Greek letters, a practice that Linnaeus also espoused. As such, "*β Sphaeria confluens*" was a subspecies of *Sphaeria albicans*.[70] In the same volume, however, he interchanged the two concepts and described several individual plants as a "variety *or* subspecies," leaving future readers to squint through his "unfortunate vagueness."[71] By the publication of his *Synopsis plantarum* (1805), Persoon proclaimed assuredly that a subspecies indicated "major morphological variations," and varieties only "minor ones."[72] Botanists of the twentieth century would later complain that he set botanical taxonomy back a century.[73]

Discourse on variation underwent a marked shift with the release of Charles Darwin's (1809–1882) *On the Origin of Species* in 1859. By then, varieties had sparked confusion among the naturalist community, and thinking of variation among individual types led to thinking about variations of populations. Pre-Darwinian natural history saw variation as "imperfections of ideal God-given 'types.'" After Darwin, this view shifted by aggregating individual characteristics into populations with shared variegations.[74] Darwin admitted that one would be "at first much perplexed" to determine what exactly constituted a species or a variety. No "clear line of demarcation" differentiated a species from a subspecies or a subspecies from a variety or even between varieties and their "lesser differences." Such lesser differences, he wrote, were "the forms which in the opinion of some naturalists come very near to, but do not quite arrive at, the rank of a species." Instead, he opined, these differences "blend into each other in an insensible series." Because the matter was well up for debate, for Darwin, the "well-marked variety" could be called an "incipient species," but the terms would never achieve full certainty. "Species," in his view, was "arbitrarily given, for the sake of convenience, to a set of individuals closely resembling each other." This application did not "essentially differ from the term variety," applied to "more fluctuating forms" and also used "arbitrarily, and for mere convenience' sake."[75] Even as the excitement of Darwin's work reached US shores, botanists such as Asa Gray (1810–1888) could only remark, with a tinge of fatigue, "There are varieties and varieties."[76] Gray preferred to throw out the subspecies term altogether, instead considering species with patent morphological deviations as varieties.[77]

Soon after the publication of *On the Origin of Species*, communities of naturalists began to reconsider the natural world using a revised classificatory scheme that reflected developments in phylogenetics.[78] Darwinian findings journeyed along axes of exchange that would also come to mark the second wave of imperial expansion, paving the way for forms of interimperial cooperation through Europe and across the pond.[79]

After Napoleon's defeat in 1815, European powers began to increasingly step into other states' affairs "in the name of humanitarian ideals and civilization."[80] Though the notion of "natural rights"—the bedrock of the idea of humanitarianism—had existed in religious, political, and juridical European discourse for centuries, as Davide Rodogno asserts, "national and transnational humanitarianism originated in the politics and philosophy of eighteenth-century liberalism." Into the nineteenth century, European governments' humanitarian interventions were grounded in the creeds of empire—namely, the expansion of civilization and the wiping out of "barbarism."[81] This common philosophy under the cloak of transnational relations likewise refashioned imperial science and the project of colonial scientization. Waving the banner of shared scientific standards within the virtue of collaboration, European botanists convened the first meeting of IBC in 1864. Following the IBC's founding, naturalists would go on to form international bodies well into the twentieth century, including the International Geographical Union (first convened in 1871) and maiden conferences representing ornithologists (1884), entomologists (1910), and astronomers (1919).

For European botanists, one of the most pressing joint concerns was fine-tuning nomenclatural practices—a longstanding perturbation in the Linnaean effort to order the world's barbarous plant names and regulate botanists' own seeming inconsistencies across their own vernaculars. Amid botany's new imperial intensification, French-Swiss botanist Alphonse Pyramus de Candolle (1806–1893), son of Augustin, served as the IBC's first president. Under his leadership, the IBC adopted its first laws of nomenclature in 1867.[82] For the IBC, nomenclature had a two-part function: "to avoid or reject" names that could generate "error or ambiguity, or throw confusion into science," and to circumvent "any useless introduction of new names." While directing the IBC, Candolle quibbled, "You scruple to create a genus? Make a subgenus or a section. You hesitate about making a species? Let it be a subspecies, or a variety. . . . By this means a multitude of new names, above all of species, that would be contested are avoided." He then explained that a variety differed by its "less striking" characteristics: "a slighter degree in character and in heredity constitute divisions of varieties or subvarieties." Keeping a variety as such, in his estimation, would keep any species determination from being contaminated by even more erroneous names.[83]

Ahead of the fourth congress meeting in Paris, Candolle produced a sixty-page primer on nomenclatural rules for the IBC to consider. A continental team of botanists joined him to amend the proposal before the meeting: Barthélemy Charles Joseph Du Mortier (1797–1878) of Belgium; August Wilhelm Eichler (1839–1887) of Hesse; and from France, Hugh Algernon Weddell (1819–1877), Ernest Saint-Charles Cosson (1819–1889), Jules Émile Planchon (1823–1888), and Louis Édouard Bureau (1830–1918).[84] At the fourth congress, Candolle and his colleagues sought to rectify the "instability of a nomenclature" and the alarming "increase of names proceeding from the different views taken of genera and species." "Natural History," they determined, "can make no real progress without a regular system of nomenclature, acknowledged and used by a large majority of naturalists of all countries." They alleged that horticulturists in particular were also "eager to get out of the chaos that they have themselves created in the nomenclature of cultivated varieties." Varieties were to be distinguishable within a single species. Scientific names were expected to be in Latin or with "as great a resemblance as possible to the original Latin names." Moreover, botanists were not to create generic names from "barbarous tongues, unless those names be frequently quoted in books of travel, and have an agreeable form that adapts itself readily to the Latin tongue, and to the tongues of civilized countries."[85]

No botanist representing a Spanish institution attended the first meetings of the IBC, with the exception of a Valencian horticulturist who joined the 1865 proceedings. This absence is not surprising considering the political tumult prior to, and immediately following, the Glorious Revolution. Intellectual repression on the peninsula during Isabella II's reign may have impeded more Spanish participation, as intense censorship and religious conservatism marked her administration.[86] Following her overthrow, after Colmeiro assumed the RJB directorship, a post he held until his death in 1901, he along with other metropolitan intellectuals founded the Sociedad Española de Historia Natural in 1871 and its accompanying publication, *Anales de la Sociedad Española de Historia Natural* (Annals of the Spanish Society of National History). By 1900, Spain had at least one IBC member, Francisco Ghersi y Vila, and several attendees.[87]

In his own work on Iberian flora published in 1885, Colmeiro touched on the issue of plant identification, referring to "mere varieties" as a regular problem confronting naturalists attempting to distinguish plants with slight variation. He went on to remark on systematics, a field which by then had seen another massive shift with the publication of *Genera plantarum* (1862–1883) by English botanists George Bentham (1800–1884) and Joseph Dalton Hooker (1817–1911). Bentham and Hooker's updated system for plant arrangement no sooner appeared than was soon replaced by the system put forth by German botanists Adolf Engler (1844–1930) and Karl Prantl (1849–1893), which reflected Darwinian phylogenetic developments, published from 1887 through 1915. Inheriting, in my reading, a skepticism apparent in Adanson's writings, Colmeiro cautioned that any "linear series," as systematics was wont to create, for organizing plant life would be "defective": "the most perfect system for some," he wrote, "does not fulfill the aspirations of others."[88]

The Case of Rice in the Philippines

Under the internationalizing ambit of botany, the problem of the variety begs the question: for whom were varieties a problem? A sampling from the concept's European intellectual history demonstrates the problem bedeviled those who were invested in systematic arrangement, in structuring plant life along connected paths of relation, and in a shared nomenclatural tongue. Banu Subramaniam observes that a "fundamentally recursive structure" emerges in which the concept of variation surfaces in different scientific disciplines, different time periods, and different political contexts. For Subramaniam, the topic

FIGURE 1.5. "Cosecha del palay" (Rice harvest, ca. 1850). Illustrator Juan Serapio T. Nepomuceno depicts a busy agricultural scene: farmers have just reaped piles of mature rice panicles; others are preparing to thresh seeds from their chaff. Fowl, carabao, and a dog contribute to the scene's activity. Viewers recognize that the affair takes many hands and hooves. Reproduction permission courtesy of the Filipinas Heritage Library, Ayala Museum.

of variation during Darwin's time and onward set the stage for population studies that informed the growth of eugenics, among other fields. The variety preceded studies of *difference* that scientists and social policymakers harnessed, deploying generalizations over populations in the interest of manufacturing fit and ideal societies, such that "the benign language of variation [was] thus converted into the profoundly political language of difference."[89]

In botany, the "variety" emerged as a term for forms that ever so slightly departed from the systematic stringency of European systematics. As the previous section shows, frustration, concession, and lassitude emerge across the sources and suggest a nagging persistence of the question that would irritate not only communities impacted by the idea of variation but also botanists in their recursive attempt to find exactitude amid the vague. Whether through the work of plant life itself or through human mediation, variation would not submit to stable theoretical conceptualization.

In this presumed insensibility, rice taxonomies in the Philippines rather begin to make more sense and exhibit a taxonomic confidence for what would otherwise be considered "varieties." For Linnaeus, "color, smell, taste, hairiness, curliness, fullness, and deformity are mostly variable and rarely constant."[90] A historical look at Philippine rice, however, opens the granary door to the socioculturally constructed sensory metrics used to distinguish varieties that challenged European systematics.

Scholars have examined how scientific researchers historically transformed "undisciplined" sensory judgments into "objective" truths unmitigated by, among other factors, aesthetic preferences, mood, and quotidian environments.[91] In early seventeenth-century England, as the influx of plants from imperial exploration flooded apothecaries and cabinets, printed herbals—several long reliant on gustatory descriptions—became the site of what would be considered intellectual inconsistencies. "Unstable categories of knowledge" motivated trained physicians to implement standards of medicinal description that attempted to do away with "inherently subjective" parameters of measurement.[92] Sensing, or a spectrum of measuring to feeling, became converted from a "time dependent (subjective) condition" into a "time-independent (objective) footprint."[93] A kind of timeless consistency then emerges within the history of the botanical taxonomic ideal, as in the case of the JBM catalogs, a text-artifact of systematic botany attempting to commensurate local rice varieties with *O. sativa*.

In one Tagalog dictionary, bontot pusa or cat's tail, translated as a genus of rice perhaps in reference to the appearance of the variety's stalk, or itlog balang or locust eggs, as in the possible shape of the rice seed, existed at one point in the Philippines among some informants as significant rice varieties.[94] Sinao may have been called such because of its pearlescent appearance, smooth and clear upon hulling.[95] Others were types of rice known for their rate of maturity: bulacan took seven months, while bunlay was fast-maturing.[96] *Violacea*, recognized for its distinct taste and its purple grains, also went by the Tagalog name tangi, which could imply a number of associations in its archaic verb form "to divide" or "to crack with one's hands or feet," or in its adjectival form, as something special.[97] In Bicol, bulao got its name from its red or copper-like husk, the word itself also referring to red hair or red animals.[98] In Tagalog, binayoyo was both a species of rice and also a kind of tree. Balir, too, was a variety of rice as well as a leaf of wild betel.[99]

A notion of time emerges among the Philippine varieties: one must sit (un)comfortably with the possibility that these names were idiosyncratic to the speaker, partially reflective of "consistent" rice forms, and perhaps as changeable— or as predictably reasonable—as the seasons or from generation to generation of speaker or from plot to plot of rice field. Unlike *O. sativa*, named by Linnaeus in 1753 in *Species plantarum*, the varieties upend the "sensibility" of European systematics. Different worlds of knowing emerge within each name, each evading full comprehension by the unwitting reader. When recalling those who shared such names for García to catalog, one obtains a sense that "*Oryza sativa*" does not sit supreme as the nomenclatural umbrella for all varieties. Instead, each variety existed within time frames and geographic scales significant to the people—and momentarily, for the record-keeper—who engaged with them.

Read for a dynamic disjuncture, the JBM seed catalogs present a refined list while pointing to a broader philosophical challenge in botany. They convey a Philippine local's confident mediation of a much longer tradition of epistemic jostling in European natural history and systematics. For the foreign reader, they present *O. sativa* based on the arrangement, yet further reveal sundry vocabularies for rice that, to those aware, reflect larger social and interspecies dynamics of agriculture and cuisine. That said, the patchwork of records reviewed thus far also implies that informants were interviewed for various purposes, systematic and lexicographic. These partial lists and snippets of information do not provide the deepest sense of the taxonomies in their full expression and practice.

As John Dupré writes, "The vocabularies of the timber merchant, the furrier, or even the herbalist may involve subtle distinctions between types of organisms; there is no obligation that these distinctions coincide with those of the taxonomist." To remedy what may be a perceptual problem of "being" and "naming," he proposes the concept of "promiscuous realism": "realism derives from the fact that there are many sameness relations that serve to distinguish classes of organisms in ways that are relevant to various concerns; the promiscuity derives from the fact that none of these relations is privileged."[100] Inspired by Subramaniam's and Dupré's observations, I recast the variety problem of European plant systematics as a problem of the asymptotic taxonomy. The paradigm of Linnaeus-inspired systematics presented a mode of thinking that would attempt a description of the perfect order of the natural world but would never quite reach it—an impossibility of which naturalists were well aware. The problem inherent in this aspiration was that the model system imagined as stable actually was not. Through the eighteenth and nineteenth centuries, neither the Anglo-European botanist nor his intellectual protégés could divine an enduring or stable system surrounding varieties. The Philippine-based interlocutors, who worked with botanists, naturalists, and lexicographers, did not seem to approach rice in the same way as the imperial systematists: their evident taxonomies did not strive toward an asymptotic ideal. García's work to record rice names under *O. sativa* may have established commensurability between the varieties and the species itself. Yet the absence of Latin equivalents for varietal names showed how such names could exceed taxonomic order, and that within each varietal name, there would be ways of knowing inaccessible to the traditional systematist.

The catalogs capture European botanical systematics of the nineteenth century that approached but never fully arrived at the assemblages of names associated with rice. The two systems, however, are not diametrically opposed—botanical order vis-à-vis the "bedlam" of varieties. Rather, their relation reveals one system's attempt to effect certainty over taxonomies that were highly locally dependent. At the start of the twentieth century, García suggested as much when reflecting on his own rice research. By then, he was the leading Manila-based expert on Philippine rice varieties, having served as a rice curator at three European expositions: the Exposición de las Filipinas in Madrid (1887), where his exhibit "Palay (Oriza sativa)" of 144 varieties of rice was handsomely awarded, the Exposición Universal in Barcelona (1888), and the Exposition Universelle in Paris (1889).[101] Writing in 1905, García noted varieties and their unique names in the southern Luzon and southern Tagalog regions as different from those found in Pangasinan, Nueva Ecija, and Tarlac, as well as those in Zambales. These varieties, in his view, could be divided regionally and by their growing patterns. He explained, "Scientific division [of rice] made by botanists includes many species, varieties, and subvarieties," implying that the "scientific" categories for *O. sativa* offered but one way individuals could make sense of rice.[102]

* * *

The mid-nineteenth century saw the development of the first secular botanical garden in Manila. The product of a partnership between the government and a private entity, the JBM began as a modest institution. Its two principal employees, Espejo and García, followed metropolitan directives to enhance the garden's scientific appeal, an inclination that had been informed by increasing liberalism on the peninsula that revitalized botany practice. The two began to produce seed catalogs reflecting the JBM's growing collection and their botanical acumen. García's aptitude, in particular, in listing local varieties of rice within their catalogs pointed to a much wider epistemological problem with which European naturalists had been contending. Varieties plagued the exactitude by which botanists from Linnaeus to Candolle hoped to taxonomically identify and systematically arrange plants. Standardizing identifications and arrangement became one of the foremost objectives of the IBC, a newly formed assembly working within the telos of empire, but the dozens of rice varieties that lived within and beyond the JBM catalogs demonstrated worlds of knowing that could outpace Latinizing schemes. Even if Latin binomials could be projected as the canopy above all other varieties, the immensity of varietal forms and those reliant on them reminded Europeans that their efforts were, in several respects, asymptotic.

Contemporary genetic techniques continue the complicated scientific debate on varieties. For what might be considered the "same" species, such techniques can demonstrate altogether genetically unique species (or new phylogenetic trees) even if, by following the classical definition of a species, they can mate with one another. As Lincoln Taiz elaborates, "Human-made categories of species, subspecies, and varieties come down to genetic variation, but precisely where we draw the lines between species and subspecies, and between subspecies and varieties is completely arbitrary. This is why the whole problem of accurately assigning plants to these categories is not just difficult, it's impossible." According to him, local, indigenous knowledge on plant varieties has been shown to correspond to subtle genetic variability, offering hints that help scientists sort through the genetic precursors that lead to different phenotypic traits.[103]

In the years after the first published catalogs, rice experienced a major commercial shift. The Philippines exported rice considerably until the 1870s.[104] Yet by the early 1870s and in the years following, the colony relied substantially on rice imports from Saigon, among other locations.[105] This dependence was due, in part, to the rededication of Philippine rice plots to the cultivation of exportable commodities such as sugar and abacá, as opposed to other edible staples.[106] Other entrepreneurial activities took precedence over rice production alongside population growth that put a strain on the domestic rice market. By the 1880s, Hokkien Chinese and European merchants controlled the trade, and by then the colony had become a net importer of the grain.[107]

Receiving little praise for the work, García and Espejo failed to garner funding from the peninsula to enhance the JBM's operations. By the 1870s, the Spanish state clearly struggled to develop its own robust agricultural economy in the Philippines through the garden's infrastructure. The "mad quagmire" that had incited the desire for a beautification project had become a swampy target of sarcasm.[108] Dejected and diagnosed with herpetic sores in 1873, Espejo returned to the peninsula for a sabbatical that never ended.[109] García remained at the garden after Espejo's departure and continued to oversee its limited functions despite what would become its ongoing turnover of Spanish personnel. The pivot to botany in the Philippines, however, had already set in motion another set of institutional and intellectual shifts. This was a key point in how colonial botanists would seek to position the Philippines on the imperial stage.

2

Scientific Statecraft

> Regino García was a Filipino distinguished by his knowledge of botany and of the flora of the Philippines.
> —TRINIDAD H. PARDO DE TAVERA, "Dibujos de plantas por Regino García," 1920

At some point in the 1870s, Regino García put pencil to paper to illustrate *Anisoptera thurifera* (see figure 2.1), a tree species listed in Manuel Blanco's *Flora de Filipinas*. García worked as the lead illustrator for the publication's magisterial reissue. He practiced the image carefully, mindful of precisely depicting the species' inflorescence, its panicle orientation, an exemplary flower, and its sepals. The bottom of the sketch reads, "*Anisoptera thurifera* Bl.," a nomenclatural correction to what Blanco had first identified as *Dipterocarpus thurifer*. The

FIGURE 2.1. "*Anisoptera thurifera* Blanco," in "Dibujos de plantas por Regino García," no. 82, ca. 1870s. Reproduction permission courtesy of Pardo de Tavera and Special Collections, Rizal Library, Ateneo de Manila University.

sketch, one of dozens prepared for the *Flora*, paled in pictorial and systematic complexity to the larger art project that lay ahead for García and for Spanish colonial botany in the Philippines.

During his earliest years in government employ, García worked at the Jardín Botánico de Manila (JBM) by day and perfected his skill in watercolors and oils by night.[1] He had scant formal training in botany prior to his start at the institution. Still, he flourished as he spent years at the Spanish outfit. Later in his career, his expertise at the intersection of art and modern botany would be most clearly expressed in 1883 through the publication of *Sinopsis de familias y generos de plantas leñosas de Filipinas* (Synopsis of families and genera of Philippine flowering plants; hereafter *Sinopsis*). The project was different from Blanco's *Flora*. The illustrated atlas of the *Sinopsis*, García's solo work, departed from the visual conventions of Anglo-European systematics. The totality of the project, nonetheless, served as a critical marker for escalated botany operations in Manila and for the metropole's pursuit of scientific statecraft.

Scientific Self-Fashioning

As Resil B. Mojares writes, "No decade in Philippine intellectual history... [was]... as productive and as consequential as the 1880s."[2] He cites several ilustrado books from the decade, including that of José Rizal, politician Pedro Paterno (1857–1911), journalist and folklorist Isabelo de los Reyes (1864–1938), and Trinidad Pardo de Tavera. To this shelf, I add *Sinopsis*. Written by Catalan botanist Sebastián Vidal y Soler and with García serving as its illustrator and lithographer, *Sinopsis* captures late Spanish colonial botany in the Philippines at its apex. A project of the Comisión de la Flora y Estadística Forestal de Filipinas (Philippine Flora and Forestry Commission; hereafter Comisión), the work broadcasted the botanical research conducted on the archipelago. At the same time, its illustrations presented a visual approach to natural life unseen in contemporary synopses, reflecting a sovereign vernacular of botanical representation original to the milieu of intellectual production of the late nineteenth-century Philippines. Botany, as I show, had a place in Mojares's Filipino Enlightenment through the hand of García. With *Sinopsis*, I argue, Iberian officials brokered intellectual exchange with rising European empires to advance a scientific statecraft. Philippine botany gave Madrid diplomatic teeth, enabling the metropole to fashion for itself intellectual might and liberality.

I use the term "scientific statecraft" knowing full well that the state-science nexus is not newfangled. That science could be harnessed to serve the interests of imperial states has been well documented.[3] Nonetheless, science had a more explicitly central role in the imperialism of the late nineteenth century. Imperial expansion in Africa, Asia, and Oceania occasioned demand for novel tools to survey and study new colonial territories: geology, forestry, acclimatization, and medicine were just a few of the fields that experienced a resurgence.[4] Increased international collaboration ensured that science could serve as the currency by which empires could build their own knowledge stores.[5] This attribute, I contend, differentiates the late nineteenth century from other periods in the history of Spanish colonial science in the Philippines. The scientific statecraft in the last decades of Spanish rule in the archipelago was especially outward facing and, as I show in this chapter, such statecraft relied on the intellectual ingenuity of those working in the Pacific colony.

García has an important place in this history. He and Vidal collaborated often during Vidal's appointment in the Philippines, embarking together on collecting missions, publications, and multinational tours. The two have been remembered as close associates, if not dear friends.[6] The Catalan has been memorialized as one of the most productive Spanish-era colonial botanists of the time, a view partially reinforced in Anglophone scholarship by US colonial assessments of Spanish botany.[7] Following US professional standards of the turn of the century, the fact that Vidal trained in systematics, published, and traveled to compare Philippine plant specimens with those in the "master" herbarium collections of Europe overshadowed other scientific labor.[8] His impactful writings and activities notwithstanding, his professional excellence should also be understood in relation to his Philippine-born colleagues. Their expertise would lend, in García's case, a visual grammar by which Vidal's work would gain repute such that Philippine botany could be instrumentalized for imperial ends.

This chapter begins with a profile of Vidal, whose deployment reinvigorated colonial botany. His reports, translation work, and itinerancy as a representative of insular operations cloaked Philippine botany in international appeal. From there, I turn to García's development as a botanical illustrator and plant systematist. The atmosphere of elite Manila fostered a community of fine artists, several of whom were also employed in insular operations. Working in the late nineteenth century, these artists followed a centuries-long foreign naturalist tradition of visualizing Philippine flora. García began to deploy his arts acumen, which would lead, among other projects, to *Sinopsis*. I conduct a formal analysis of his atlas, comprising 100 lithograph plates depicting roughly 1,900 plant figures. I compare his illustrations to others he completed on behalf of the Augustinian-backed reissue of *Flora de Filipinas* and other illustrated natural science synopses of the time. In doing so, I demonstrate how *Sinopsis* enacted a novel visual logic that speaks to the eruption of native intellectual activity, presenting García's slight retreat from other foreign visions of local plant life. The work itself mobilized Philippine plants through his artistic techniques and systematizing choices. An invitation for heightened imperial scientific investigation, I argue, it furthered imperial undertakings in the floristic region known as Malesia.

Since the mid-nineteenth century, metropolitan and colonial botanists shared botanical data, specimens, and illustrations that contributed to the phytogeographic study and discursive emergence of Malesia. By the mid-twentieth century, botanists understood the region to include the Dutch East Indies, the Malay Peninsula, Sarawak, British North Borneo, Brunei, the Philippines, Christmas Island, Portuguese Timor, New Guinea, and the Bismarck archipelago.[9] When European naturalists observed common flora that crossed colonial boundaries and distinguished the region from others nearby, they realized the need to work concertedly to produce a more comprehensive botanical snapshot of this large geographical expanse. The momentum of imperial consolidation propelled these intellectual moves.[10] In the case of *Sinopsis*, routes of intellectual exchange served Spain and other empires looking to capitalize on the promise of colonial expansion in Malesia.

An Institutional Restyling

Thirteen years into the JBM's existence, Madrid shifted control of the garden to the Inspección General de Montes (IGM), the Spanish scientific forestry unit in the Philippines that had become fully functional in 1863.[11] While Philippine forests had long transfixed the earliest foreign inhabitants of the archipelago, Spanish authorities worked more methodically to develop the colony's forest economy from the late eighteenth century onward.[12] Upon the IGM's founding, Iberian botanists, agronomists, and foresters arrived to the Philippines to survey all manner of plants from the commercially scalable to the increasingly rare. This was coincident with a greater overseas effort to conserve timber-bearing forests in colonial Puerto Rico and Cuba, which Iberian and colonial officials believed were threatened by unfettered consumption and unsuitable weather patterns.[13] The IGM's responsibilities included land surveying, issuing licenses for tree felling, and policing what was deemed illegal logging.[14] As one of its foremost employees once described, though the "*terra ignota* can lend itself to expansive flights of fantasy," Philippine land contained a "precious treasure that can be the source of very significant income for the state."[15]

The end of the lucrative galleon trade in 1815 led to increased commercialization and urbanization in Manila through the nineteenth century. Foreign and local entrepreneurs migrated to the capital to seize newly opened trade opportunities. The influx of merchants and workers from neighboring provinces heightened the demand for urban infrastructure, and Manila became "the ultimate destination for wood."[16] The first IGM employees contributed to the production of biannual reports that covered timber assessments in places such as Tarlac and on expeditions to the mountains of Makiling, San Cristobal, San Pablo, and Banajao.[17] IGM operations met both the commercial demands of Manila and the bureaucratic demands of Iberian officials.

Following Zoilo Espejo's departure from the JBM, the colonial government appointed a pharmacist with a keen interest in medicinal plants to serve as the garden's interim director.[18] Officials in Madrid overrode this appointment, voicing greater confidence in IGM employees to recalibrate the institution's purpose.[19] Vidal was fundamental to this goal of transforming the rigor of Spanish scientific operations.

In 1871, the Ministerio de Ultramar selected Vidal, who had trained previously in Madrid, Tharandt, and Zurich, to be the chief engineer of the Philippine IGM.[20] Once in Manila in 1872, Vidal likely met García, yet they did not set out immediately to reform the garden. Instead, Vidal surveyed forests and appraised timber-bearing trees for commercial exploitation, implementing a two-pronged approach to the Philippine environment by combining forestry and botany. In his *Memoria sobre el ramo de montes en las islas Filipinas* (Report on Philippine forests, 1874; hereafter *Memoria*), he advocated the tactical combination of the two related fields to ensure the most lucrative outcomes for the state. He bemoaned what he considered a lack of systematic evaluations, herbarium-grade samples, and taxonomic identifications. In his view, an engineer with training in botany could more readily distinguish "the species that rise gigantically before him, trunk half-hidden under lianas that embrace it... orchids in their fantastic shapes and brilliant hues, the whimsical fern fronds and the wrapping of their foliage among the leaves of a hundred climbers." Without botany, an untrained observer would simply be confused.[21]

Vidal suffered an accidental point-blank gunshot wound while in Tayabas in 1873. He returned to the peninsula to recover and undertook a lengthy project to translate *Reisen in den Philippinen* (Travels in the Philippines, 1873) by ethnologist and naturalist Fedor Jagor (1816–1900). Vidal's training in Tharandt and Zurich presumably gave him the confidence to attempt the translation.[22] *Reisen* chronicled Jagor's travels through parts of Luzon and the Visayas in 1859 through 1860 and included commentary on Philippine agricultural products, environmental phenomena and formations, and Spanish colonial-bureaucratic and clerical systems. In the translation's opening, Vidal lauded Jagor's work for its "scientific accuracy and precision" and hoped translation into Spanish could stimulate travel to and study of the Philippines.[23]

Jagor's tract on the Philippines was distinctly colonial in outlook, even if not completed on the terrain of a formal German colony. *Reisen*'s ethnological assessment of the archipelago reiterated racial hierarchies set against a landscape of valuable natural resources. The work also offered a sober appraisal of the Spanish colonial government and boasted more complete and accurate data than any Spanish findings. Leading up to the second wave of European expansion, as Nathaniel Parker Weston has argued, German naturalists "forecast[ed] the actualization of the Imperial German nation-state in 1871 and its subsequent establishment of overseas colonies in Africa, Asia, and Oceania" through natural history work on the Philippines. Such a move articulated the superiority of a German Empire yet to come.[24]

After the translation was published in 1875, Governor-General of the Philippines José Malcampo y Monge (1828–1880) demanded its censorship, insisting, "The book overflows with ideas contrary to patriotic interests, ideas whose propagation is of utmost necessity to prevent among the different races that populate this country."[25] Whether Malcampo successfully enacted his resolution is unclear: the work was listed as essential reading on the Philippines, and at least one state-run trade school's library carried it into the 1890s.[26] In Manila and in Madrid, no sources suggest that Vidal was professionally reprimanded for his translation. In fact, Vidal regularly engaged with German scholarship, which showed an openness to extra-imperial information on the Philippines that could enhance his scientific-ambassadorial position in the succeeding decade and a half.[27] Translation enabled both an old empire and a budding one to agree upon a shared lexicon of territorial and intellectual conquest.

Territorial and intellectual collaboration was necessary to execute proper regional studies of Malesia, first defined as a unique floristic region in 1857 by Swiss botanist Heinrich Zollinger (1818–1859). A floristic region implied a commonality of plant families across a geographical zone, and Zollinger argued that Malesia defied the boundaries of British, Dutch, Spanish, and Portuguese colonial territories.[28] Specializing in Malesian flora, therefore, required interimperial interaction. A Spanish botanist could execute a broad study of Philippine flora, and a Dutch botanist that of the East Indies, but indexing a family of plants that spanned Malesia required engaging with Spanish, Dutch, and British botanical tracts, if not more.

This development in botanical science corresponded with enriched relations between Spain and Germany. Whereas late nineteenth-century Spanish and German scholarship has been characterized as the "arena of competition and rivalry" between the two European powers, politicking beyond printed scholarly pages showed the contrary.[29] Soon after Vidal submitted his *Reisen* translation to the Spanish state, he received third-class honors from the Prussian Crown and formal accolades from the Ministry of Agriculture in Berlin—not for his translation but for *Memoria*, which he donated to the library of the Prussian Ministry of Agriculture.[30] In light of Vidal's awards, Malcampo's objections more intensely juxtapose Madrid's sense of scientific statecraft, which had a distinctly interimperial dimension. Vidal's work thus initiated a series of exchanges that continued through his career and highlighted Philippine botany as a diplomatic instrument to which *Sinopsis* would be central.

The Ministerio de Ultramar designated Vidal the head of the newly created Comisión in 1876 following his return to the Philippines.[31] Blueprinted in the *Memoria*, the Comisión was a specialized body dedicated to botanical study of the Philippines.[32] The ministry voiced the intrinsic link between state wealth and the natural sciences to rationalize the body's formation. The Comisión set out to enact Vidal's two-pronged strategy: first through statistical surveys, and second through a systematic classification. According to officials, the work that the religious orders completed on Philippine plants—while informative and pioneering—was neither methodical nor internationally comparative in scope. In the eyes of the metropole, IGM engineers were more intellectually equipped for such tasks.[33]

Immediately following the Comisión's establishment, Vidal departed Manila again. International travel was a privilege, and among colonial scientific personnel, Vidal had the exceptional benefit of travel that even more senior functionaries could not readily enjoy.[34] This distinction, I surmise, signals the value he would continue to represent for the interests of the Spanish state. Upon his departure, he acted as secretary of the Comisión de Ultramar in Philadelphia for the Centennial International Exhibition in 1876, a US exposition of industry, natural history, agriculture, and the arts modeled after the European tradition of universal exhibitions.[35] The ministry notified the governors-general of the Philippines, Cuba, and Puerto Rico of the appointment, suggesting that Vidal's growing approbation in Madrid heightened his standing over much of the similar scientific work in Spain's remaining colonies. As he left for Philadelphia in 1876, plans to initiate the Comisión and the hiring of its personnel unfolded over two years.[36]

Visualizing Philippine Plants

The Comisión's primary tasks reflected standard European botany practice. Under the terms outlined by Vidal and the Ministerio de Ultramar, the Comisión had to describe phanerogamic (seed-bearing) and vascular cryptogamic (non-seed-bearing) species and to cross-reference findings with available data. Published descriptions of new species and their structures required accompanying illustrations suitable enough to advance study and classification.[37] During the rollout of the Comisión, García, as one of two active horticulturists at the JBM, was appointed as a forestry assistant.[38] With his institutional memory, he oversaw the construction of a research pavilion at the JBM that enabled the Comisión's classificatory and herbarium work.[39] The Comisión also named Cayetano Argüelles y Fernández (b. ca. 1845) as its natural history conservator, and Francisco Domingo y Casas (b. ca. 1845) as its draftsman. All three appointees would steadily visualize Philippine flora.[40]

García, Argüelles, and Domingo have been described as an "uncommon breed of artists-naturalists," who were simultaneously academic painters and employees of the Comisión.[41] Their combined capacities may have been rare for the archipelago, but for aristocratic Manila, fine arts and natural history were tightly braided by the second half of the nineteenth century. The cosmopolitan capital with its lively neighborhoods and bustling port provided its wealthiest residents opportunities for social advancement unmatched elsewhere. A growing middle class in Manila and its neighboring provinces increased the number of students and demand for literacy education and specialized training.[42] The high social standing of many well-heeled native and mestizo families ensured young students' entry into the established secondary schools of the city.

In 1865, the Spanish Crown standardized and required secondary education in the Philippines, which increased access to higher education and to vocational work in the industrial arts, nautical science, drawing, and at the JBM's Escuela de Agricultura, land surveying, botany, and agriculture. In its earliest iteration, the Escuela de Agricultura maintained programs for "obreros alumnos," or worker-students, with and without scholarships to complete training. The specific racial and class composition of the school's student body remains unclear. To obtain a scholarship, however, required a "healthy and robust complexion" and passing the school's entrance exams. To obtain higher government-appointed ranks within the institution, candidates needed written and reading fluency in Spanish, skills often reserved for the elite across the islands even after the Crown's promulgation.[43]

The oldest child of a well-to-do family, García was primed to study land surveying. His parents sent him to the Escuela Náutica, originally established as a merchant shipping school, in the walled city of Intramuros. Similar to other visual artists of the time who had briefly enrolled in the school, García then trained at the Academia del Dibujo y Pintura (ADP).[44] Debate exists as to precise founding of the ADP. The Real Sociedad Económica (RSE), the same body that developed proposals for the JBM, founded the institution around 1821 to 1823.⁴⁵ Then known as the Academia de Dibujo, the RSE likely merged its academy with the private art school of Tondo-born painter Damián Domingo y Gabor (1796–1834) that had been founded anywhere from two to eight years earlier.⁴⁶ Domingo boasted of the institution's open acceptance of Spanish, mestizo, and native enrollees.⁴⁷ Lack of funds led to the ADP's closing in 1834, but its patrons reestablished it in 1845.⁴⁸

The Spanish academic tradition, specifically that of the Escuela de Bellas Artes de San Fernando, influenced the instruction at the ADP. Copies of Iberian works, many of them religious or royal in subject, served as models for ADP students to duplicate as they trained. Baroque paintings by Bartolomé Esteban Murillo (1618–1682), Jusepe de Ribera (1591–1652), Alonso Cano Almansa (1601–1667), and Guido Reni (1575–1642) set in place the institution's artistic foundation. According to Concha Díaz Pascual, such a selection shows the metropole's impression of the Philippine colony's artistic tastes and the visual style to which its students were expected to aspire.⁴⁹ The "baroque" style, as has been elaborated by José Antontio Maravall and applied by Patrick D. Flores to the Philippine context, was more than just an artistic period with aesthetic influence but a feudal orientation that had characterized much of archipelago's relations with Spain and the Spanish colonizing habitus.⁵⁰ Prescriptive as this may have been, students' sensibilities departed from the purely baroque and could have been molded by what cosmopolitan Manila had to offer. By the mid-nineteenth century, privately owned presses ended religious orders' long control of publishing.⁵¹ New serials circulated iconography, mercantile imagery, and handmade illustrations that were not devotional and could transmit artistic ideas more up-to-date than what was taught at the ADP. The flood of capital molded the art market and how ADP artists could value their own work as part of the domestic and global political economy.⁵² The academy was the only institution that sent government-funded students to train in the metropole. By the 1890s, however, the government granted the opportunity to only a little more than a dozen pensionados. In 1891 alone, the ADP had approximately 540 enrollees, gesturing to the even greater number of students that developed visual styles shaped by an insular education.⁵³ "Philippine academic art," as Luciano P.R. Santiago terms it, may have been informed by the peninsula but ADP artists worked beyond its limits.⁵⁴

As one such colony-bound student, García painted provincial scenes and presented native subjects. In his 1877 oil, "El labrador" (see figure 2.2), a barefoot farmer walks aside a carabao in a rural setting. The low horizon intimates a relatively flat plane, likely for agriculture. Structures stand to the left of the

FIGURE 2.2. "El labrador" (The farmer) by Regino García y Baza, 1877. Oil on canvas, 90 × 140 cm, © Photographic Archive Museo Nacional del Prado. Reproduction permission courtesy of the Museo Nacional del Prado.

frame, faintly showing thatch architecture and fencing. For a scene in which the farmer is nominally central, the focal point is shared by a native buffalo, the human figure donning a salakot, and his local plow. Different flora surround them, distinguished enough by type to suggest a botanically diverse environment. Light hits the human figure's back, only slightly illuminating his face, to suggest sunrise or sunset. The painting departs from the stunning contrasts of Baroque pieces with García's choice of shadows alongside the vast pastoral sky. Its level of detail, perhaps best executed in the man's shirt, evidences more formal training in figure drawing that he would have obtained at the ADP. The painting's sophistication assembles a fuller, more realistic scene, filled to the edges of the canvas with the local environment, than was customary in the tipos del país genre, which literally endeavored to represent different types of people in the Philippines and in which the earliest ADP students excelled. Nonetheless, "El labrador" does not take on antique, courtly, or historical subject matter. What's more, it is noticeably simpler than the busier activity seen in peninsular costumbrismo. The staid farmer juxtaposes the boisterousness of the locals' tavern or of the wet market, quietly en route to or from work.

Akin to García, artists lent their dual expertise in illustration and natural history to Spanish scientific operations. Despite being appointed as a plant

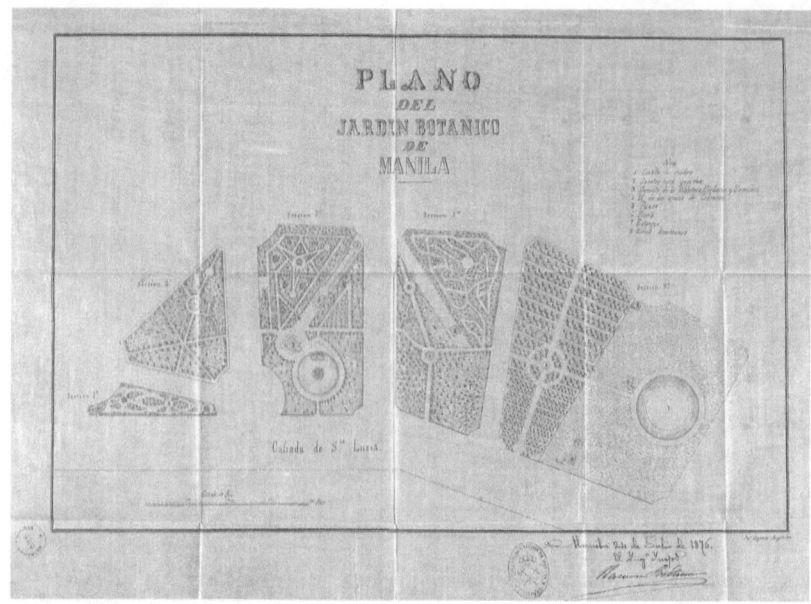

FIGURE 2.3. "Plano del Jardín Botánico de Manila" (Plan of the JBM) by Cayetano Argüelles, approved by Ramón Jordana on July 24, 1876. The numbered key locates structures on the JBM grounds, including a library, herbarium, pond, and guard houses. Ultramar, MPD.5489. Reproduction permission courtesy of the Archivo Histórico Nacional.

conservator, Argüelles completed illustrations on behalf of the IGM, including institutional layouts (see figure 2.3). He specialized in Philippine woods and, for part of his career, was stationed in Laguna under the IGM auspices.[55]

Both Argüelles and Francisco Domingo, the grandson of Damián, along with a cadre of Philippine- and peninsula-born artists, many of whom likely practiced in the halls of the ADP or in association with its professors, went on to execute illustrations for the Comisión and the IGM. Recruiting this class of painters changed the dynamic of Philippine botany, putting at the IGM and Comisión's disposal an arsenal of visualizing capacity founded in Spanish academic conventions yet wrought by the local specificities of art production.

Illustrations of Philippine flora had by no means been absent earlier in the Spanish colonial period. The seventeenth century witnessed a number of illustrated natural histories by Catholic missionaries. The work of Bohemian Jesuit Georg Joseph Kamel (1661–1706) originally contained images, but these were unpublished, and their absence likely led to his corpus' fall to obscurity.[56] Similarly, Ignacio de Mercado Morales (ca. 1648–1698) wrote and illustrated a manuscript on Philippine medicinal plants, but few copies of the work reportedly survived.[57] Francisco Ignacio Alcina's (1610–1674) multivolume account of the Visayas (1668) also contained plant illustrations.

Artists recruited for the Malaspina Expedition (1789–1794) produced hundreds of images of Philippine flora, yet the manuscripts, folios, and data gathered from the expedition in the late eighteenth century saw little public light. The secretive Spanish state, then protective of its maritime holdings and cognizant of its imperial rivals' interests, stifled the circulation of such natural history knowledge, especially in published formats.[58] The expedition under Juan José Ruperto de Cuéllar y Villanueba (ca. 1739–1801), who sought commercially promising plants on behalf of Spain and private business interests in the Philippines, had a similar outcome.[59] Philippine-born artists José Lodén, Tomás Nazario, and Miguel de los Reyes collaborated with Cuéllar, who remitted eighty images of Philippine plants back to Madrid.[60] These images did not travel much farther.[61]

Raquel A. G. Reyes has speculated that had the work of secular botanists on the Philippine leg of the Malaspina Expedition been published, "the authority of clerical writings would surely have been challenged long before the late nineteenth century."[62] Up through the mid-nineteenth century, missionaries produced the manuscripts and publications many are familiar with today on early modern Philippine plants. These documents, likely due to cost, did not voyage, even if they feature renditions of species or the tropical landscape.[63] Blanco's *Flora de Filipinas* and its posthumously produced second edition had no illustrations. Not until its update in the 1870s and 1880s did several of the artists associated with the Comisión illustrate its pages. That team also included the Manila-born painter Emina Jackson y Zaragoza (b. ca. 1858), the spouse of IGM engineer Domingo Vidal y Soler (1838–1878), the editor of the reissue and Sebastián's older brother. Women of her social position either attended one of the few women-only schools or gained training by exposure to cultural networks in the capital. The ADP did not admit women until 1889.[64] Jackson herself enjoyed her early education abroad, enrolling in schools in London and Paris.[65] Similar to the other painters, she likely frequented the sphere of artists-naturalists of the time.[66]

Sinopsis: A Research Invitation

Illustrated works offered a "visual and verbal vocabulary" that could be shared over greater distances.[67] Newly foraged and preserved plant specimens had their limitations. IGM employees regularly remitted plant material to Manila, and Vidal and the Comisión remitted herbarium sheets to the Real Jardín Botánico in Madrid. Yet illustrated publications could be mobilized more readily than a perishable live material or a fragile, dried specimen on an herbarium sheet—and could visually transport more structures of a typical plant.[68] A single specimen's image, for instance, might show each stage of the specimen's reproductive cycle, even though the constraints of time and the organic development of a species would impede its faithful collection at each reproductive stage. Its visual representation would also likely be the composite of several specimen samples from which the illustrated type would be created. Even if herbarium samples were the most preferred material for botanical investigation, illustrations could better weather the uncertainties of travel, insect infestation, and rot that plagued plant samples in transit from a colonial place.[69] Such efficiency was an aim of *Sinopsis*.

Vidal's preface to *Sinopsis* positioned the work within the collective empiricism of late nineteenth-century European botany.[70] In it, he declared that "in order not to fall to an eclecticism that could lead to complications," the *Sinopsis* was a compilation that conformed to what he viewed as an "epoch" in modern science, Bentham and Hooker systematics. Vidal acquired the first two volumes of Bentham and Hooker's *Genera plantarum*, among other German and British botanical works, for Philippine operations while touring on behalf of the Comisión de Ultramar in London in 1877.[71] That Joseph Hooker had invited Vidal to use the Kew facilities at his disposal might explain some of the flourish with which he wrote of the English botanists.[72] Echoing Carl Linnaeus's writings of the century prior, the "eclecticism" he denounced likely referred to a legacy of publications that failed to adhere strictly to one taxonomic standard or systematic arrangement. Ordering *Sinopsis* along a single—and at that point, the most recently devised—system could brandish the publication's intellectual capital. Vidal further framed *Sinopsis* as a guide for future research expeditions: an invitation for follow-up studies not limited to Spanish botanists or the Spanish Empire alone. The index to García's atlas listed collecting localities to "facilitate verifications" upon which newer studies could be built. The atlas, celebrated as "the most important part" of the publication, served as the central element of this invitation toward more scientific work in the Philippines.[73]

Instead of single specimens, the plates in *Sinopsis* demonstrate species and representative plant structures of entire plant families. The synopsis genre in the natural sciences typically explored shared characteristics among members of the same taxonomic family. García's visual presentation of plant material, however, was unusual. First, not all synopses flaunted plant illustrations, such as the 646-page *Synopsis analytique de la flore des environs de Paris destiné aux herborisations* (Analytical synopsis of the flora surrounding Paris intended for

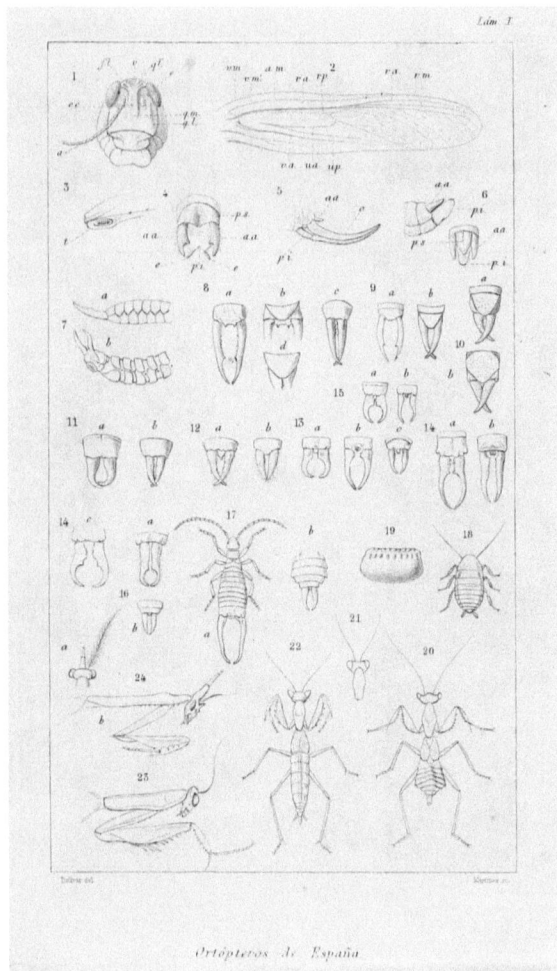

FIGURE 2.4.
Plate 1 in *Sinópsis de los ortópteros de España y Portugal* by Ignacio Bolívar y Urrutia (Madrid: T. Fortanet, 1876–78). Image courtesy of Smithsonian Libraries and Archives.

botanical study, third edition 1876) by Ernest Saint-Charles Cosson. Second, García's style was uncommon for Spanish colonial botany in the late nineteenth century: no other colonial contemporary of *Sinopsis* offers such a display. On the peninsula, two major synopses immediately preceded Vidal and García's: *Sinópsis de los ortópteros de España y Portugal* (Synopsis of orthoptera in Spain and Portugal; 1876) by Ignacio Bolívar y Urrutia (1850–1944) and *Sinópsis de las especies fósiles que se han encontrado en España* (Synopsis of fossil species found in Spain, 1878) by Lucas Mallada y Pueyo (1841–1921). Bolívar's was an etymological survey with only seven illustrated plates (see figure 2.4), and Mallada's work on faunal and floral fossils had only thirty-six. The plates in Bolívar's and Mallada's publications also present structures arranged in a relatively linear horizontal

SCIENTIFIC STATECRAFT 69

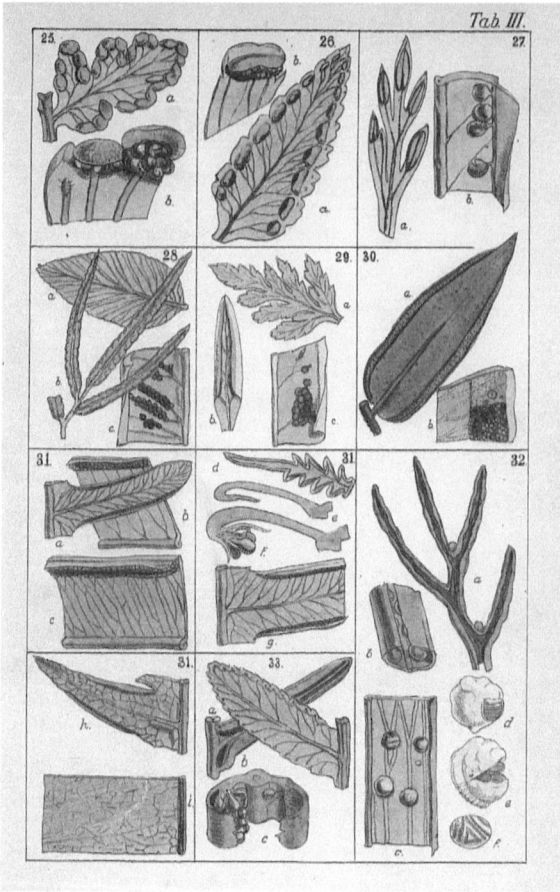

FIGURE 2.5. Table 3 in *Synopsis filicum* by William Jackson Hooker and John Gilbert Baker (London: R. Hardwicke, 1874). Image courtesy of the LuEsther T. Mertz Library of the New York Botanical Garden.

or vertical fashion, like the presentation of plants in *Synopsis filicum* (1874) by English botanists William Jackson Hooker (1785–1865) and John Gilbert Baker (1834–1920). Fern structures are boxed and segmented by genus (see figure 2.5).

In *Sinopsis*, García altered such visual organization. He captured plants' reproductive stages, combined several species reported as members of the same family of plants, and did so with less linearity of arrangement. His plate for the Euphorbiaceae family (see figure 2.6), for example, presents the structures of species from eight different genera. Letters on the plate indicate species while numbers distinguish species' structures, requiring more careful scrutiny because of the absence of linearity. In figure 2.7, "A" refers to *Mallotus moluccanus* Müll., or the kukui nut tree. García separated structures to illuminate parts of the plant, like its fruiting branch, flower, fruit, and fruit cross-section. He centered the *M. moluccanus* branch (A-1) on the page, allowing its enlarged

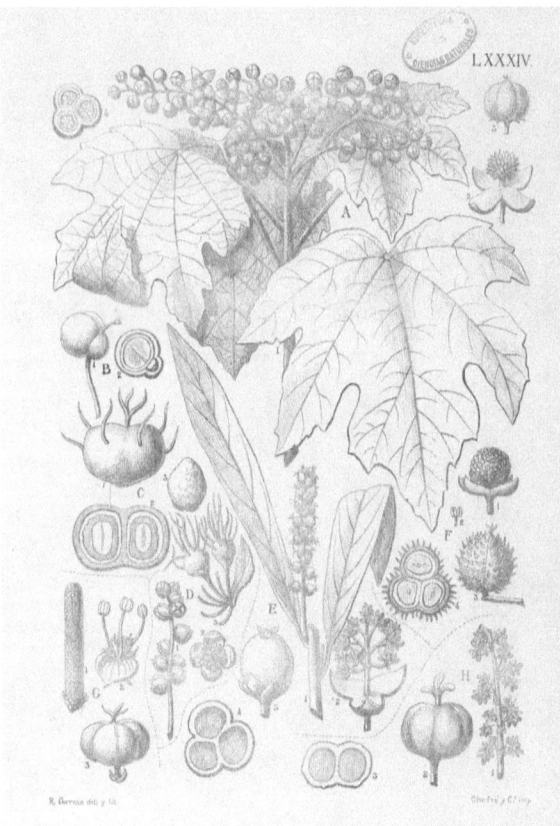

FIGURE 2.6. Plate 84, "*Euforbiáceas*," in *Sinopsis*, vol. 2, atlas (Manila: Chofré y Compañía, 1883). The plate features the structures of eight distinct species presented as members of the same taxonomic family. Image courtesy of Biblioteca Digital of the Real Jardín Botánico.

leaves, including one that has been detached to offer full view of the leaf's midrib and veins, to take focus. He also provided the inflorescences of other species at a much smaller scale (D-1, G-1, and H-1). For species B through H, García favored their fruits, flowers, and seeds, and sorted each species with faint dashed lines. These elements largely compose the bottom half of the plate. Through these sorts of structures, he endeavored to show what would have been considered part of the reproductive stages of the varied specimens. In the case of the *Macaranga mappa*, he even provided a sample of a fertilized flower (D-3).

Many of the plates in *Sinopsis* contain an unusual abundance of plant material crowded to the edges of each page, perhaps for financial reasons. Printing an illustrated survey of Philippine plants was costly. The publication of the five-hundred-page *Bosquejo geográfico é historico-natural del archipiélago filipino* (Geographic and natural history sketch of the Philippine archipelago, 1885) by Philippine IGM forester Ramón Jordana y Morera (1839–1900) required at least 1,500 pesos for printing in Madrid and had only twelve lithograph plates.[74]

SCIENTIFIC STATECRAFT 71

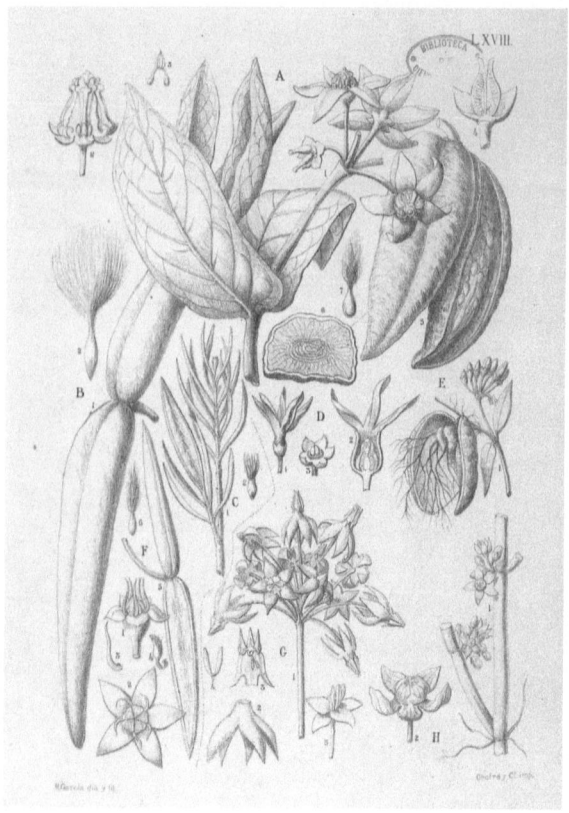

FIGURE 2.7. Plate 68, "*Asclepiadeas*," in *Sinopsis*, vol. 2, atlas (Manila: Chofré y Compañía, 1883). Image courtesy of Biblioteca Digital of the Real Jardín Botánico.

This sum was equivalent to 150 percent of the total budget for materials in the Comisión's first year of operations.[75]

Vidal and García likely chose a local publisher for purposes of proximity and cost. The Manila-based publisher of *Sinopsis*, Chofré y Compañía, released children's publications, civil reports, and other documents of the Comisión along with significant intellectual works like de los Reyes's *El Folk-lore filipino* (1889).[76] The publishing house provided commercial printing services from 1882 through 1898, and its Spanish proprietor, Salvador Chofré, imported a lithographic machine to the colonial capital.[77] With lithography at the Comisión's disposal, Vidal and García could produce more copies of *Sinopsis* and its atlas at a faster rate. By October 1883, at least 106 of the 250 copies planned for peninsular distribution had been sent to libraries, research centers, specialized schools, and press houses.[78] García's illustration style could have contributed to the overall lower cost of *Sinopsis*, which may have helped make the work more accessible. Printed with no color, *Sinopsis* was not only cheaper to print but also cheaper to buy. Compared to the opulent four-volume revision of Blanco's *Flora de Filipinas*, *Sinopsis* illustrated more individual structures and presented botanical classification in a manner unseen in the Augustinian work.

Flora de Filipinas has been lauded as "the crowning glory of Philippine art and science in the colonial era" because, in my reading, of its colored images. In 1876, the Order of Saint Augustine in the Philippines initiated its reissue. While the text of the revised editions was printed in Manila, the 477 lithographs were printed in Barcelona. Five hundred colored copies were produced, with another thousand printed with black-and-white plates. The Augustinians paid 73,000 pesos for the entirety of the project—a sum that was more than double the personnel costs of the Comisión in its total eight years of operation.[79]

Flora de Filipinas featured Iberian- and Philippine-born painters in Manila, including García's brothers Rosendo (b. ca. 1848), a pharmacy graduate who specialized in medicinal plants, and Juán (b. ca. 1850).[80] García signed thirty-five plates for *Flora*, though he is believed to have painted all the unsigned plates in the reissue. The majority of *Flora* illustrators trained at the ADP under the peninsular Agustín Sáez y Glanadell (1828–1891), who joined the creole Lorenzo Rocha é Icaza (1838–1898), the only other professor at the ADP, to judge the submissions for the reissue's frontispiece.[81] Under their directorship, ADP students worked with oils, Contè crayons (a waxier alternative to pastels, derived from pigments, clay, and graphite), varnishes, charcoal, Gilbert pencils, and a variety of paper media, among other tools.[82] *Flora*'s visual conventions likely emanated from the ADP's standard of training and the expectations of the Augustinian financiers. Rocha himself trained at the ADP and enrolled as the first Philippine-born pensionado at the Escuela de Bellas Artes in 1858.[83] The Escuela de Bellas Artes had a history of turning out skilled, well-paid artists who could produce natural history illustrations.[84] Due to the ADP's artistic consultation for the project and given the backgrounds of most of its contributors, Santiago suggests viewing the work as a "'Who's Who'" of Manila's fine arts society. Further, the work demonstrates the type of gatekeeping at play to preserve the visual manner of botanical illustration.[85]

Following this line of elite production, *Flora de Filipinas* was prohibitively expensive for individuals. Its cost is especially evident in the grand style of the illustrations and the presentation of plant material on its pages. The *Flora* plates, like so many others in the oeuvre of natural history illustration, conforms to a visibilizing and invisibilizing norm, as Daniela Bleichmar characterizes it: single specimens are made visible against a "sea of white" while specimens' places of origin are summarily expunged.[86] When compared with the images

FIGURE 2.8. "*Ocimum americanum:—Blanco*," *Flora de Filipinas*, lámina 407, vol. 2 (Barcelona: Verdaguer, 1883). Image courtesy of Biblioteca Digital Hispánica, Biblioteca Nacional de España.

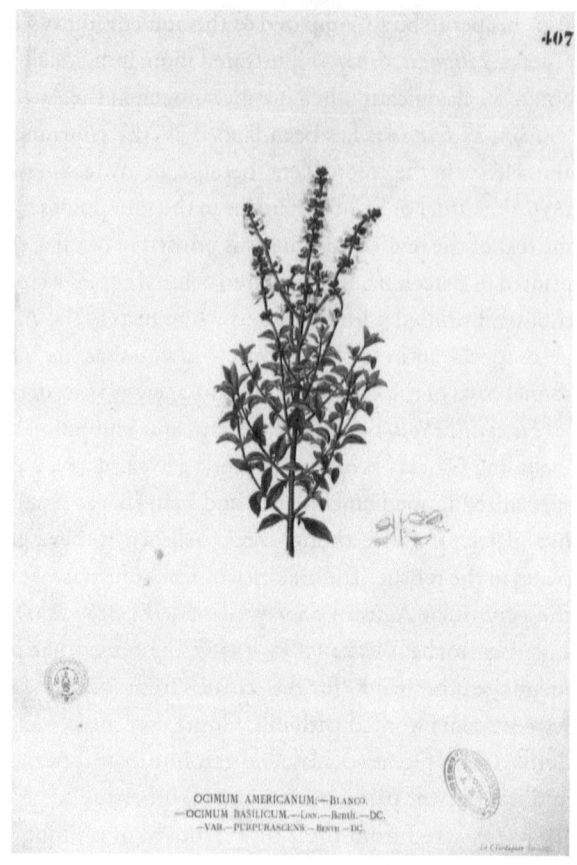

in *Sinopsis*, the plates in *Flora* reveal the opulence of negative space and color, as evident in the plate for *Ocimum americanum* (see figure 2.8). Centered on the page is a flowering branch of the species, resplendent with leaves in shades of green and blossoming buds of gentle pink. The illustration is dynamic: while some flowers have fully bloomed, others only partially so. The same pink and brown hues that comprise the flowers are used to highlight the underside of the sample's leaves and provide visual balance to the image. To the specimen's left is a faint sketch of what looks like a portion of a stem with leaves. Incomplete and uncolored, the portion offers the total plate a sense of being a work-in-progress toward the production of the focal point, the colored *O. americanum*. Apparently faithful to the size of the plant, the illustrator—likely García—did not magnify the structure, leaving much more expensive blank space unused.

To identify the locality and any previous publication on the species, *Flora* readers needed to purchase the illustrated volume alongside the text. This manner

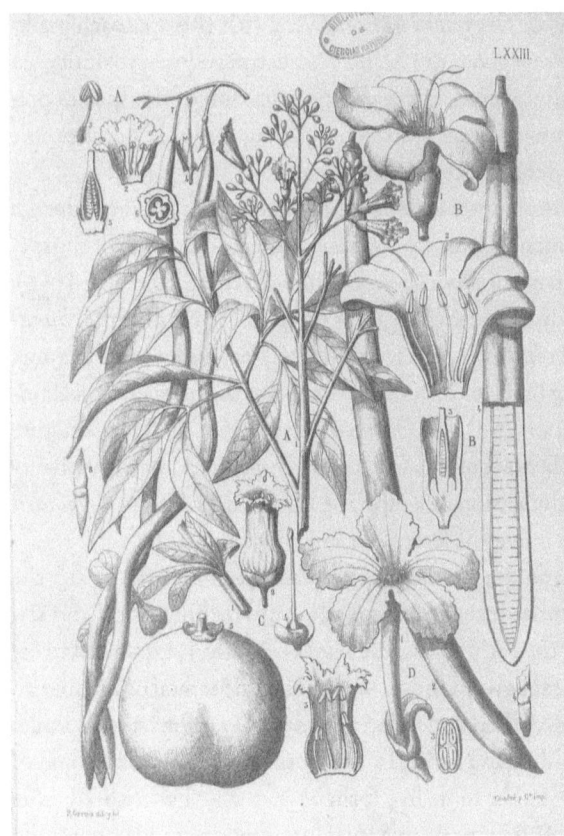

FIGURE 2.9. Plate 73, "Bignoniáceas," in *Sinopsis*, vol. 2, atlas (Manila: Chofré y Compañía, 1883). This plate features the structures of four distinct species of the same taxonomic family. Image courtesy of Biblioteca Digital of the Real Jardín Botánico.

of publication, in which four volumes constituted a complete set, reinforces the project's affluence. Like *Flora*, the *Sinopsis* has a separate atlas, except the atlas's index also functions as a key for the plates, indicates specimens' collecting localities, and includes additional collecting data. The index also references corresponding page numbers in the *Sinopsis* text, suggesting that the volumes should be read together.

Even in the absence of color and negative space, García's plates for *Sinopsis* showcase considerable technical skill to present plant structures in service to plant classification. Although cost may have contributed to the visual production, García's compositional capacity made the atlas a work of modern colonial botany and complex artistry. The plate for the Bignoniaceae family (see figure 2.9), for instance, presents four different genera and species. Unlike the previous *Sinopsis* plate seen in figure 2.7, rather than using faint dashed lines of separation, he differentiated the structures by repeating letters, like A and B, to indicate that certain numbered structures correspond to species such as *Stereospermum*

SCIENTIFIC STATECRAFT 75

quadripinnatum Fern. (A) or *Oroxylum indicum* Vent. (B). This approach works to avoid confusing observers given the A and B structures' close proximity. At the same time, the visually layered, entangled presentation of the species presents a plate far more complicated in its totality than a simple floral profile conventional in other styles of representation.

To illustrate *O. indicum*'s open corolla and stamens (B-2), García drew a cross-section of the specimen's flower that overlays a longitudinal slice of the species' fruit (B-4) and the fruit of *Dolichandrone rheedh* Benth. & Hook. (D-2), which itself extends nearly the entire length of the plate. The fruit for *S. quadripinnatum* (A-7) also extends the full length of the plate and weaves through leafed (C-1) and flowering (A-1) branches. The flowers of the plate take focal primacy, like that of *O. indicum* (B-1) and *D. rheedh* (D-1), since no other structure overlays them. They cast shadows on the structures behind them, thereby enhancing the illusion of three-dimensional depth. The flowering branch of *S. quadripinnatum* (A-1) does the same, as its leaves cast shadows on its fruit (A-7).

This profusion of plant specimens on the plate gestures less toward the limited finances of the entire project and more toward technical skill and scientific advancement. By the early 1880s, the IGM's operations expanded to include outlying provinces far beyond those surrounding the capital of Manila. To complete *Sinopsis*, officials carried out collecting work in locations such as Iloilo, Tanay, Bataan, Tarlac, and Nueva Ecija. In the work, Vidal also cited some of the JBM's own material—either in its live form or that which was stored in its seed bank or herbarium. This textual and visual information could counter the negative appraisal that had beleaguered the garden early in its establishment.

Crowded plates and finished indices maximize a sense of reinvigorated research and local ingenuity. In place of just one species, the plates present systematic relations accentuated by the original aesthetic taste that García likely employed at his own discretion. Working on his own artistically, he took creative liberties that played with size, scale, and positioning that would have required intense precision to draw on lithography stones. Because of their lively, entangled composition, García's plates trenchantly depart from the pictorial norms of synopses of the time. Their wound, gnarled, or layered presentation makes some of the plates less easily discernible, more organically vivacious than the boxed or plainly laid structures found in *Sinópsis de los ortópteros* or *Synopsis filicum*.

Such an emergent visual style has been observed in other colonial contexts. In Bleichmar's study of José Celestino Mutis's (1732–1808) New Granada Expedition (1783–1816), Mutis's workshop of some sixty American plant illustrators produced a distinctive style that, while in adherence to the European botanical illustration genre, brandished a taste for symmetry and visual flatness (as opposed to "volumetric naturalism") unseen in typical European works. Mutis maintained a careful watch over his workshop, engaging iteratively with artists, who evidenced their own artistic inclinations and skills, to ensure accuracy and consistency of style.[87] García's rather independently completed project, likely done in consultation with Vidal's knowledge of systematics, means he did not have to conform to various parties' expectations, such as those that dictated the *Flora*. In some respects, the collective and highly coordinated manner in which the *Flora* was completed shares more similarities with Mutis's project. Nevertheless, a simultaneous visual making and unmaking of botany appears in the atlas to *Sinopsis*, manifested by García's understanding of the science and of artistic innovation. Notions of systematic affinity still restrict what viewers may see of the Philippine environment. Structures are divorced from ecological and sociocultural context in service to efficiency, inheriting a historically imperial mode with which to render colonial natures.[88]

Artistically speaking, the plates depart from a more two-dimensional approach that had characterized earlier Philippine botanical drawings, such as those attributed to Lodén, Nazario, and de los Reyes. The two-dimensional style itself was a Western introduction to the Philippines, copied in the sixteenth and seventeenth centuries from European images.[89] By illustrating the structural resemblances within a single family, García presented an updated approach to colonial botany that, until that point, had either not circulated as comprehensive set of images or had ventured to provide only single-species illustrations. His command of the plant life emerges then within ilustrado production, combining European disciplinary elements alongside an artistic sensibility idiosyncratic to his familiarity with the environment.[90] As Trinidad H. Pardo de Tavera would later write of him, García was indeed a "Filipino" in the emerging political sense of the term as one "of the Philippines" who, like his ilustrados peers, could produce self-consciously competent work on the islands.[91] García had knowledge of botany *and* of Philippine flora, combining both through his own sovereign vernacular.

Toward a Scientific Statecraft

While the colored plates of *Flora de Filipinas* present the gems of Philippine plant life, their single-species presentation was not meant to spur family-wide investigations within what was projected to be the Malesian plant world. This kind of encouragement of increased study of the regional tropical environment coincided with the publication of *Sinopsis*. Instead of keeping it cloistered in

Iberian and colonial Spanish libraries, the Spanish state shared *Sinopsis* with other imperial plant specialists. Even Vidal did not see the work's scope as limited to Spain and the Philippines. In addition to the other writings of the Comisión, he promoted the *Sinopsis* as an achievement that could stand with the work of British, Dutch, and German botanists, especially that which had been conducted on "the great Malay archipelago" or Malesia.⁹²

By the mid-1880s, Philippine botany consistently drew more international acclaim from European botanists engaged in Malesian study. In 1885, the Italian plenipotentiary in Spain appealed to the Ministro de Estado for a copy of the illustrated *Flora* on behalf of Odoardo Beccari (1843–1920), a Malesian palm specialist. Between 1877 and 1889, Beccari published a three-volume study on Malesian flora drawn from thirteen years of fieldwork in the region. The plenipotentiary communicated Beccari's delight if he were to obtain Blanco's work from the Spanish government. In return, he offered Beccari's *Malesia*. Beccari also wanted "as complete as possible [a set] of Philippine plants from the director of the Jardín Botánico de Manila," in exchange for a collection of Malesian and Papuasian plants provided by himself. The plenipotentiary emphasized that such a trade would render a great service to science.⁹³

Whether Beccari had a professional relationship with Vidal or García remains unclear. While the record shows that they had professional ties to the same botanists at Kew, no available metropolitan documents reveal direct correspondence between them. By the time of the plenipotentiary's correspondence, however, not all work of the JBM or the Comisión had to course through peninsular channels. Unlike the earliest years of JBM operations, Philippine samples were no longer remitted to the peninsula ahead of their disbursement to other research centers. The work of the Comisión, in particular, operated more independently as Vidal developed direct professional relationships with botanists in England and the Netherlands.⁹⁴ Indeed, Madrid was not always a centripetal force. In the broader eighteenth-century Spanish Empire, the objectives of colonial naturalists and those of the metropole at times diverged. Some colony-bound naturalists of the Philippines and the Americas tipped the intellectual center of gravity toward their local concerns and regional preoccupations, displacing Madrid as the only fount of scientific directive.⁹⁵ By the late nineteenth century, colonial researchers in the Philippines expanded their own networks beyond metropolitan Iberia to include other imperial nodes. At this point in late Spanish-era colonial botany, the Comisión, JBM, and IGM functioned with a "complex autonomy," as seen at the British colonial garden in Singapore, and did not have to fully defer to the whims of the metropole.⁹⁶ After receiving a copy of the Italian communiqué, the Ministerio de Ultramar relayed to the colonial government in Manila that any exchange of plants was at Vidal's discretion.[97]

The Ministerio de Ultramar, however, denied Beccari's wish for *Flora*. It clarified that *Flora* was not a work of the Spanish state but, instead, of the Augustinian Order. In its place, the ministry offered *Sinopsis* as another work—one funded by the coffers of the Spanish state—that was "of no less import that could be very useful" to Beccari.[98]

Suggesting that the Spanish state stifled the imperial circulation of *Flora* would be inaccurate (it was, after all, a very expensive set of volumes), yet its promoting *Sinopsis* as a product of the secular state is telling. In some respects, this move reflected the political tenor of the time in the metropole and in the colony. With Isabella II's deposal and the surge of liberalism on the peninsula came the appointment of Philippine Governor-General Carlos María de la Torre (1809–1879), who has been remembered as a more liberal, well-received colonial functionary. Rocha, one of the artistic overseers of the *Flora*, closely aligned with de la Torre and the liberal excitement of the time.[99] The governor-general issued a number of progressive measures, including a push to secularize the clergy of Manila's University of Santo Tomas. Yet, with the ascension of the conservative Prince Amadeo of Savoy (1845–1890) to the Spanish Crown in 1870, the reactionary Rafael de Izquierdo (1820–1883) replaced de la Torre.[100] Izquierdo's government oversaw the execution of those associated with the Cavite Mutiny.

In 1872, troops protesting increased colonial taxation staged a revolt in the province of Cavite. Izquierdo's forces suppressed the uprising, which served as a pretext for the arrest of known liberals and the execution of prominent native secular priests, who by then comprised a significant political formation. The events of 1872 implicated among other colonial cruelties those mediated by Catholic friars. By the late nineteenth century, native clergy had self-organized into a more formidable power against friar orders, which had "a *de facto* monopoly" over parishes in the Philippines. Because Spanish colonial governance proliferated through churches, parishes had been the heart of civic life throughout much of archipelago.[101] The mutiny has been seen as one of the events that triggered the reformist and revolutionary movements in the Philippines wherein anti-friar rhetoric spread in the colony and on the peninsula. The Spanish state's decision to distinguish its intellectual work from the Augustinian publication is therefore curious, considering how the state may have been responsive to these liberal developments.

In other respects, Spanish statecraft could instrumentalize scientific developments in the Philippines and those by its colonial functionaries, proudly maintaining them as the product of government moneys, no matter how paltry.

Spain leveraged Philippine botanical work to present itself to Germany and Italy as a revived intellectual authority. Praise for *Sinopsis* continued to run imperial axes during a series of international expositions through the 1880s. In addition to the Philadelphia exposition of 1876, Vidal served as a representative of metropolitan and colonial interests at the Internationale Koloniale en Uitvoerhandel Tentoonstelling in Amsterdam in 1883, the same year *Sinopsis* was published. From his meeting with other botanists in the Netherlands, Vidal oversaw the delivery of more coveted Malesian plant material to Dutch repositories.[102] Ahead of his return to Manila in 1884, the Dutch state made Vidal a Knight of the Order of the Lion of the Netherlands, likely for his work on the Comisión and for the specimens.[103] Within weeks, the Spanish government also commended both Vidal and García for their publication of *Sinopsis*. The state awarded Vidal the Encomienda de Número de Isabel Católica and García, the Cross of Carlos III.[104] In the end, Beccari appreciatively provided copies of *Malesia* in exchange for *Sinopsis*.[105]

Vidal built the Comisión's botanical storehouse once back in the Philippines. From his collected numbers, he published *Revisión de plantas vasculares de Filipinas* (Revision of Philippine vascular plants; hereafter *Revisión*) in 1886, for which García again provided illustrated plates. That same year, the Comisión dissolved due to funding setbacks. Vidal and García's internationally oriented botanical work, however, did not. Among other former Comisión personnel, both were appointed as custodians of Philippine flora for Spain's 1887 Exposición de Filipinas and the Exposición universal in Barcelona in 1888.

Spanish government records detail Vidal and García's travels in Europe during their service toward the expositions.[106] In 1887, Vidal and García, along with officials of the office of the governor-general and of public works, Pedro Urtuoste, Abelardo Cuesta Cardenal, and Mariano Sánchez Villanueva, embarked for Madrid for the Exposición de Filipinas to manage the natural history exhibits.[107] Vidal oversaw the transport of live Philippine plant material to be put on display.[108] Leonor Paulí, Vidal's wife, and Felipe de los Santos, their "criado indígena" or native servant as it appears on passage records, accompanied him.[109] García oversaw the transport of plant material as well, including his exhibit of native rice varieties.[110]

The Exposición de Filipinas was one of Spain's attempts to showcase its national modernity. Exhibit administrators aimed to promote the "modern, scientific gaze" among exhibition-goers, who could witness modernity unfurl on its grounds.[111] Such modernity was an intensely racialized affair. Live human exhibits appeared beside vegetal ones, suggesting civilizational spectacle alongside exploitable potential. Among its many displays, the Exposición presented controversial live human exhibits composed of peoples from the Philippine, Marianas, and Caroline islands. A former Comisión official oversaw the exhibits, reflecting the facilitative role these scientific functionaries played in mounting such work.[112] These exhibits instilled both rage and patriotic fraternity in ilustrado fairgoers who wrote of the events.[113] Rizal and his compatriots, according to Filomeno V. Aguilar, found the exhibit "distasteful and offensive" and were "stirred by the appalling accommodation and treatment of the human exhibits—mirroring the way that Spain dealt with the whole colony."[114] At the same time, they complained that the delegation was "unrepresentative of their homeland."[115] Still, some ilustrados, including Rizal, later claimed fraternal association with the individuals on display.[116]

The twin commercial and economic imperatives undergirding the Exposición were equally clear.[117] Agricultural exhibits, commercial forest products, and García's rice varieties contributed to the show of available wealth in the Philippines. At the same time, as much as a display of natural abundance could intimate an abundance of exportable commodities, it also relayed Spanish colonial botany's systematic command of the environment. Writing on the botanical displays at the Expocisión, Vidal branded the Comisión's work as conducted on behalf of "modern culture" and in the name of the Spanish state. These descriptions could exemplify the pride of "scientific investigations in lands where the Spanish flag waves."[118] He and his team curated lush arrangements of vines, flowering specimens, and plant-derived products for both pompous presentation and investigative enthusiasm: a stroll through the remade tropical environment in Retiro Park enticed the learned and the lay. The arrangements included samples of the woods that had been Vidal's utmost concern upon his first trip to the Philippines. Molave, ipil, narra, yakal, and balete appeared, celebrated for their utility in civil and naval construction. Administrators split the botany exhibits from the explicitly anthropological, agricultural, and industrial.[119] Vidal boldly announced that all the natural history work in the Philippines could now competently be conducted by Spaniards, whose output would only complement that of the English, Dutch, and German botanists working on Malesia.[120] The grounds for imperial collaboration were increasingly imminent.

* * *

Vidal, García, and other members of their team returned to the Philippines from their European tour in mid-1888. A little more than a year later, Vidal died of cholera in Manila.[121] His maternal uncle, a Manila resident and painter, successfully lobbied the colonial government to raise a statue in Vidal's honor, one that sported an IGM uniform decorated with his international accolades. After his collaborator's death, García continued to produce work—at times without the benefit of direct authorship—using material from the herbarium of the former Comisión.[122] By the early 1890s, as a faculty member at the University of Santo Tomas, he wrote on behalf of the IGM. By then, the IGM's financial support began to ebb. Four different Iberian-born directors cycled through the leadership of the forestry unit and the JBM, some leaving on account of ill health.

García and Vidal's combined efforts fashioned a scientific character for the Spanish state and for colonial Philippine botany, as captured in the publication of *Sinopsis*. García's unique artistic style gave Philippine botany a distinctive quality unseen in previous secular undertakings. *Sinopsis* thus contributed to—and was flaunted within—the worlds of Malesian phytogeography and of international exhibitions. The publication aided the Spanish state as it projected itself as a scientific power relative to other empires and quite possibly to the longstanding clerical dominance in the Philippines. Colonial botany in the Philippines took on an international appeal, which enabled Spain to declare its intellectual might even as the empire's hold on its Pacific territory grew feeble.

While the scientific work continued, albeit in fits and starts, the politically tense decade upended the institutions. Revolutionary fighting destroyed part of the garden's grounds, and a fire in 1897 obliterated the local records, library, and collections associated with Spanish botanical work. The colonial government had been imprisoning and executing those suspected of taking part in politically subversive activities. Mounting pressure from intellectuals, laborers, clergy, and peasants to overthrow Spanish supremacy and to assert political reforms only increased. Anticolonial revolutionary societies staged armed uprisings throughout the islands. The decade portended radical change for the colony. These portents would come to include new visions of governance, education, and cultural life. Within these visions, a certain flower grew.

PART II
―――――

Science in a Place of Flux

3

Ubiquitous Sampaguita

> Sweetheart of this soul, whom I adore,
> you conjure the fragrant sampaga
> unsullied in purity, exalted in beauty
> wellspring of total ecstasy
>
> Captivating Eden,
> where lie the sweetest pleasure and delight,
> under your gaze,
> aromatic flowers in sudden bloom
>
> Opening lyrics to "Jocelynang Baliwag," ca. 1890s

"Jocelynang Baliwag," a kundiman or Tagalog love song, appealed to the lovestruck and revolutionary alike in Manila and its environs in the mid-1890s. Its composer dedicated it to Josefa Tiongson y Lara, a Bulakeña from the town of Baliwag and the subject of several kundiman at the time, suggesting her esteem among elite circles in the capital.[1] Distinct among the songs in her honor, "Jocelynang Baliwag" became the "kundiman of the revolution" against the Spanish colonial order.

At the turn of the century, Tagalog, Ilocano, and Spanish writing commonly featured poetic use of the sampaga, a shrub indigenous to the archipelago. Its clusters of small, white aromatic flowers were reportedly common in gardens. As vendors peddled garlands of buds and full blooms, their sugary scent would have caught passersby. The sampaga ornaments *Noli me tangere*, appearing in scenes with young women, with the indelible María Clara, and allegedly on the grave of the protagonist's slain father. In the opening stanzas to "Jocelynang Baliwag," Tiongson evokes likeness to the sampaga, the standard of "beauty" and "purity." Surrounded by "aromatic flowers in sudden bloom," she and her youthful prime reproduce a common organismic framework that equates human, sexed-as-female forms to flowers' charismatic qualities. The lyrics thereby evoke divine femininity through the sampaga, anchored to a particular image of plants' procreative potential.

Composer Antonio Molina (1894–1980), describing the kundiman's popularity among turn-of-the-century insurgents, asserted it "gave life and stoked their joyful memories of time spent together, resonant with the image of a beloved mother, spouse, or children."[2] Scholars have since duly examined the song for the role of the woman in the proto-national imagination.[3] Indeed, within its historical context, the song reflects a particular gendered invocation of the homeland common to Philippine nationalist history.[4] The sampaga within it, though essential to the tune's gendering and localizing elements, has evaded comparable scrutiny, yet such invocations of the flower at the time clearly summoned a particular vision of the Philippines.

The Political Crucible and a Plant

The colony entered a crossroads during the late nineteenth century when Spanish repression became more acutely evident. Philippine- and Europe-based ilustrados published damning tracts against administrative inadequacy and clerical abuses. Propagandist material traveled between these places to scrape away the veneer of foreign superiority. In Manila, rumors circulated about the *Noli* and its sequel, *El filibusterismo* (1891), two seething indictments of the political order.[5] An electric scene of mono- and bilingual serials also expanded. Manila publishing houses connected commentators, artists, and journalists while printing volumes on newly authored studies of the Philippines' past and present. In this politically tense and creatively effervescent climate, "Jocelynang Baliwag" stirred listeners with the sampaga.

In this chapter, I shift from colonial botany's institutional developments to examine a single plant at the intersection of culture, politics, and science during the twilight of Spanish colonialism in the Philippines, a time when the notion of sovereignty appeared across several registers. Presently, "sampaga" denotes any white-flowered jasmine species, and its Hispanized diminutive "sampaguita" refers to a type of jasmine with relatively small flowers known as *Jasminum sambac* or Arabian jasmine.[6] During the late Spanish colonial period, the two Luzon-centered plant names were interchangeable. Botanists today agree that the species originated from Bengal.[7] Local folklore suggests the name originated from the Tagalog phrase "sumpa kita" (I promise you) after sampaguita flowers emerged following the death of a grief-stricken princess betrothed to a valiant prince. The plant has different names throughout the archipelago: marol in Cebuano, malul in Maguindanao, and kampupot in Tagalog-Kapampangan, to share a few. The US colonial government declared it the national flower in 1934, an honor the Philippine state continues to celebrate. Contemporary political and cultural leaders extol its "hallowed" status as a ubiquitous "icon" of national culture.[8]

What made the flower iconic? Momentarily, I question the species' primordialism to the Philippine nation, a stance essential to interrogating the often complicated and conflicted appropriation of plants, landscapes, and nature for state-making projects.[9] In order to do so, I explore how elite knowledge production in the late nineteenth century converged to render an image of the flower as part of the everyday in the Philippines. Such an image was not without political consequence or incredible cultural impact.

The nineteenth century saw an uptick in literary and intellectual production. The first newspaper debuted in 1811 and more periodicals followed.[10] In Luzon, the mid-nineteenth century witnessed an increased number of petitions to publish books, which had been centralized to the archdiocese of Manila. Once approved by church and state censors, authors and readers could enjoy religious and secular material.[11] Peninsular and creole writers commanded the society of letters and "cultivated native 'disciples,'" as Resil B. Mojares explains, who transformed literary and intellectual production from "'pre-national'" to "distinctly 'national' in its ambition." From the 1880s onward, Philippine-born intellectuals refuted Spanish assimilationist colonial writing that had dismissed or failed to understand the archipelago's history, languages, industrial capacity, and even folkloric breadth.[12] They positioned their findings to advocate for reformist and anticolonial political ends, yet contradictions permeated their efforts, given they used the very disciplinary tools that often upheld fundamental inequality.[13]

A science almost exclusively practiced by colonial officials and the social elite, botany had a tenuous place in anticolonial and reformist politics. Botany personnel had a complex, at times indirect, role in this surge of activity. Regino García, for example, refused to participate in activities against the Spanish colonial state. He had married into the Roxas clan of Manila, which had been targeted by the state for sedition. With his spouse, Rufina, he had three children, one of whom reportedly took up revolutionary arms in 1896.[14] Never able to live or train abroad yet handsomely honored by the Spanish Crown, García was upwardly mobile in the colonial government service. Nevertheless, he allegedly signed the Malolos Constitution and joined the First Philippine Republic's efforts against US takeover of the islands.[15] Later, he actively assisted the transitional US colonial government—only to leave after a few years.[16] As for other ilustrados who waffled across political lines, such choices most certainly had material enticements, ideological considerations, and lasting outcomes.[17] In spite of the political ambivalence found among members of botany's scientific community, botany nonetheless provided tools with which intellectuals could confidently declare knowledge-claims over local flora and their matrices of meaning to foreign audiences. At the same time, across their sources, sovereign vernaculars of the sampaguita appear, departing from strict adherence to Linnaean systematics. These captured sensoria and connotation that existed beyond botanists' pages, aspects of the species in which locals were presented as well versed. For some ilustrados, botany could not quite articulate the flower's public recognizability. The science could be useful but only to an extent.

I begin by following the knowledge-project that fixed the sampaguita to its known Latin name. As I show, this centuries-long, intra-active process was not straightforward and included the sensory perceptions of colonial naturalists, corrections to Linnaean botany, and local informants' knowledge of plants. Human engagements configured the same presumable species, which had different material and cultural value over space and time. Through available published and visual material, I follow the "triangulation of image, text, and specimen" that led to the sampaguita's botanical identification and entry into scientific parlance into the late nineteenth century.[18] From there, I study how the flower emerged in elite production. The species' local names surfaced in lyrics, poetry, and art as well as lexicographic and botanical catalogs, which contrasted, sometimes quite explicitly, typical scientific depictions of the plant. I argue that the plant, saturated by gendered and gendering portrayals as well as upper-class notions of daily life, became part of the reformist political projects of the late nineteenth century. These renderings of the sampaguita, as I conclude, were then taken up, branded, and canonized as *national* decades into the twentieth century.

The Flower in Naturalist Discourse

Translating the Sampaga

Some of the earliest colonial writing on the sampaga appeared in the seventeenth century. Missionizing naturalists and lexicographers documented a locally well-known flowering species as part of a growing monopoly of intellectual production to codify the colony's island environments and languages. As early as 1582, the Manila Ecclesiastical Junta decreed the practice of translation as formal policy among its missionizing clerics. By 1602, the Spanish Crown mandated that all missionaries have competency in a native language of the Philippines.[19] Translating the word of God into terms that potential converts could comprehend complemented efforts to record the healing and useful plants of would-be parishioners. The entwined tasks could ensure successful conversion and survival in such alien terrain. At the same time, as clerics listed common vegetal forms, they brought their subjectivities to the plant worlds of Luzon and the Visayas, likening local flora to those on the Iberian Peninsula in order to "render the unfamiliar familiar."[20] They also equated Philippine flora to that found in Manila, using the existing nomenclature there to characterize species equivalence across the islands. These asymptotic techniques—efforts to classify a plant through affinities with others yet ever failing to find perfect equivalence—initiated ongoing work to make plants legible to both foreign audiences and those based in the colonial capital, thereby determining the plants' lasting labels.

One of the clearest examples of these techniques surfaces in Francisco Ignacio Alcina's natural history of the Visayas, in which the author noted his "more than average" fluency in Bisaya as the "key that best opens all secrets."[21] The large volume covers Visayan clothing, jewelry, and customs before extensively reviewing the most useful plants of the island region. Inhabiting the descriptive natural history genre, Alcina's narrative mode and sensory observations establish the grounds for communicating fact and cementing his authority.[22] He determined that the marol of his informants was Manila's sampaga, a reportedly ancient flower, giving readers little sense of his justification, only that perhaps the plant's "ancient" quality explained its growing range in the archipelago. His declaration establishes an equivalence between the two names tethered to an agreement over what the plant is commonly understood to be. For those based in or familiar with Manila, the congruence assists in invoking an image of the marol that is the same as the sampaga.

Alcina continued by describing the marol as "como nuestro jazmín" or "like our jasmine," further establishing congruence not only for colonial audiences but also for a wider Iberian-Hispanophone community of readers.[23] He described how potted shrubs, "very ornate with the white of the flowers," bore similarity to the jasmine observed abroad, without further explicating the jasmine itself and instead, insinuating an agreement as to what sensory stimuli constitute the jasmine form. The marol, however, is *not quite* the jasmine of common Iberian knowledge: its fragrance, he admitted, exceeds that of other species. Although Alcina casually referenced the saying, "el jasmín por el olor es mejor" ("the jasmine, for its scent, is better"), he conceded that the marol had a more profound aroma. The jasmine, nonetheless, was a necessary referent to achieve translation.

Historically speaking, clerics could not always achieve easy translations between Spanish and the languages they learned in the Philippines.[24] That the sampaga was "jasmine-like" reflects this empirical uncertainty. Met with the challenge of translating local Philippine flora, naturalist missionaries and translators wrestled with the ambiguity of the flower and the image they had to manufacture for their readers. Similarly, Juan José de Noceda (1861–1747) and Pedro de Sanlucar, when translating the Tagalog sampaga in their *Vocabulario de la lengua tagala* (Tagalog vocabulary) published in 1754, called it "a flower like the jasmine."[25] Their translational efforts to find congruence between colonial Philippine and peninsular flora were already slightly affected by the extent to which the sampaga was different and already expansive in meaning and use.

For newly arrived naturalists like Alcina, the archipelago was "so removed from the rest of the world, so distant from whatever makes news . . . as if they are not known by anyone."[26] Such an impression—one that would persist for centuries—privileged the foreigners arriving to a place that in fact had been widely known on its own terms. Noceda and Sanlucar's *Vocabulario* has been known for its incorporation of local poetry to deepen readers' comprehension of the usage of Tagalog words, including proverb verses such as "Nuti ang gumamila / nula ang sampaga" (The hibiscus flower whitens, the sampaga reddens) explicated to mean "the vile rise, the good fall."[27] In their vocabulary, the two provided verb forms for sampaga, including magsampaga or to adorn oneself in the flower, and magcacasampaga, or to have sampaga (typically in a garden) or to be among others fond of it. Casampagahan referred to all things related to the sampaga (not unlike this very chapter). Their Tagalog informants even modified the plant name to indicate relation between two or more individuals: a casampaga was someone with whom an individual left a place with sampaga, which could be interpreted as a close friend or a lover with whom fond memories have been shared.[28] That the sampaga was worn and grown—and that relation and axioms were defined through it—implies a greater versatility of use and being that well preceded Noceda and Sanlucar's observations.

From the late seventeenth century onward, more missionizing naturalists continued to place the sampaga in an even more geographically vast lattice of jasmine species. Georg Joseph Kamel cataloged what he observed of the sampaga during his botanical fieldwork in Manila and the capital's surrounding terrain. He maintained robust correspondence networks with European scholars, which ensured a circulation of botanical findings to and from the Philippines that shaped natural knowledge of the colony into the late seventeenth century.[29] In 1704 in *Philosophical Transactions*, Kamel identified at least six different jasmine species, three of which resembled the Arabian jasmine or zambach—a possible cognate, he surmised, to the Tagalog term.[30] However, he made no firm botanical identification.[31] His original manuscript to *Philosophical Transactions* included illustrations of the Philippine species, some of the earliest extant colonial images of what he observed to be the sampaga (see figure 3.1), but the journal did not publish them, and much of his visual oeuvre did not become known. Without visual media to correspond with his observations, Kamel's writing went largely unrecognized by future naturalists who eventually fixed the sampaga to a Latin name.[32]

Fastening the "sambac" Identity

Over a century after Kamel, Manuel Blanco's writing on the sampaga in *Flora de Filipinas* dominated most nineteenth-century investigations of the flower. Following naturalists before him, he claimed the sampaga to be the same as the Visayan "manul" (a probable variant of marol) and the Pampangan campopot. Moreover, his identification, which Spanish colonial botanists would eventually correct, linked the plant to a Latin name.

Blanco declared the sampaga to be the *Nyctanthes sambac*, a species already known in European botany parlance.[33] To make the identification, he looked to Carl Linnaeus's *Species plantarum*, which described the *N. sambac* as a species of the now-obsolete taxonomic class, Monandria.[34] Blanco relied heavily on European literature to devise the *Flora*, which he proudly organized following Linnaeus's sexual system of arrangement. In the 1758 edition of *Systema naturae*, Linnaeus wrote of the *N. sambac*, citing and correcting *Herbarium*

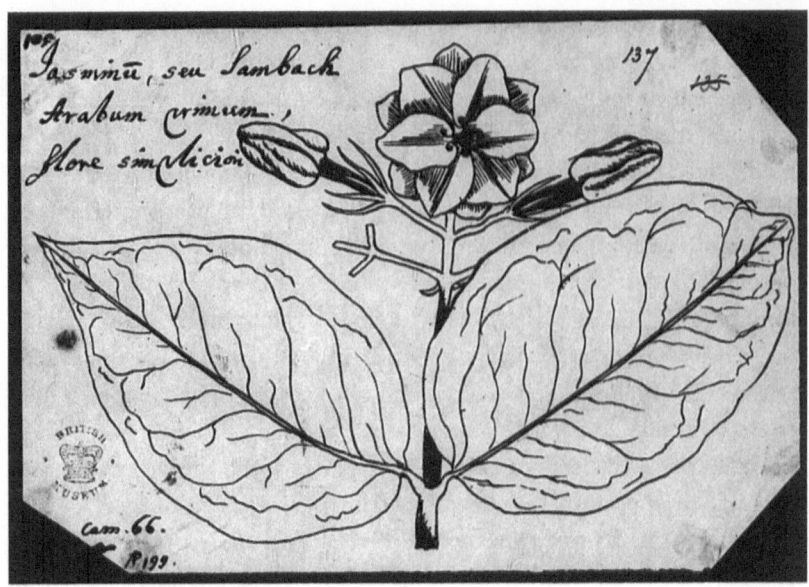

FIGURE 3.1. "Jasminum, or Sambach," one of several of Georg Joseph Kamel's unpublished illustrations of the Philippine species with purported similarity to Arabian jasmine. Sloane Manuscripts, 4081, Camel 66, R199, © British Library Board. Reproduction permission courtesy of the British Library.

amboinense (1741–1750), the posthumously published herbal by Dutch naturalist Georg Eberhard Rumphius (ca. 1627–1702). Writing from Ambon in the Dutch East Indies, Rumphius described the "most famous of all the flowers" known in Ambonese as copa puti. He considered it a jasmine but found its leaf structures distinct from other species, assigning it instead the unique name of *Flos manorae,* likely from the Malay term for the flower, bonga manoôr.[35] According to Linnaeus, Rumphius's *F. manorae* was none other than the *N. sambac* and perhaps not as unique as the Dutchman had thought.

Unbeknown to Blanco, a different European botanist disagreed with Linnaeus's determination decades after *Systemae naturae.* Scottish botanist William Aiton (1731–1793) reassigned the species from the *Nyctanthes* genus to the *Jasminum.* In his *Hortus kewensis* (1789), Aiton alleged the species to be native to the East Indies. To make the correction, he referred to Hendrik van Rheede's (1636–1691) illustrated *Hortus indicus malabaricus* (multiple volumes, 1678–1693), a flora from what is known today as the Western Ghats region of India. Van Rheede's description and illustration of the kudda-mulla included classificatory notes of its Latin name, *J. sambac.*[36] Aiton cross-referenced these observations with the writing of Swedish naturalist Daniel Carlsson Solander (1733–1782) and specimen collections at Kew to reach his conclusion, which has since been accepted by botanists.[37]

Blanco most likely did not have access to Aiton's correction and worked principally from Linnaeus. Even arriving to *N. sambac*, however, was not a linear process. Publishing Europeans aimed to achieve taxonomic certainty by branding what each of them postulated to be the same plant with a Latin name. The plant's jasmine-like days were through. Blanco relied on Linnaeus's description, which itself was based on a plant specimen likely remitted from the Indian subcontinent.[38] Linnaeus had, in turn, corrected Rumphius's illustrated observations of the same alleged species. Each node along this route was mediated by the sensory perceptions of the working naturalists, each invested in a morphological vocabulary for classifying plant life. For Linnaeus, the lower leaves of the *N. sambac* were cordate and obtuse. To Blanco, the species' stem was cylindrical and compressed; its flowers an "excellent wash for eye inflammation due to heat."[39] The leaves of the *F. manorae* were "like the leaves of the orange, but shorter and rounder," according to Rumphius.[40] This ostensibly objective vocabulary reconciled naturalists' idiosyncratic observations and value judgments, and once triangulated with illustrations and specimens, bound future descriptions of the sampaga to practices of imperial taxonomy.

Study of the sampaga went into the final decades of the century under the Comisión de la Flora y Estádistica Forestal de Filipinas. The increased personnel of the colonial body and of the Inspección General de Montes, combined with developments in steamship technology, meant that more men could travel to farther reaches of the archipelago to collect its plants. No surviving record suggests, however, that each stationed employee observed the sampaga. Foresters and surveyors regularly sent plant material to Manila, except the sampaga, likely because it was not considered to be a significant forest product.[41] Even so, Comisión personnel wrote of the ornamental flower, and an update to its Latin name appears in the revised reissue of *Flora*. The Comisión reidentified the sampaga as *J. sambac* following Aiton (see figure 3.2). The taxonomic re-identification may have been in part due to Sebastián Vidal's obtaining a copy of *Hortus kewensis* in London in 1877 for the Comisión's library in Manila.[42]

In their *Sinopsis*, Vidal and García diverged from some of Blanco's original findings in *Flora*. They distinguished the sampaga from the sampaguita: "sampagas [*sic*]" refers to the *Jasminum* genus; sampaguita, on the other hand, to *J. sambac*. Although their publication lists plant names from a number of different Philippine languages, they categorized the name sampaguita as a distinct nomenclatural form of "Manila." It comes as little surprise that the two found

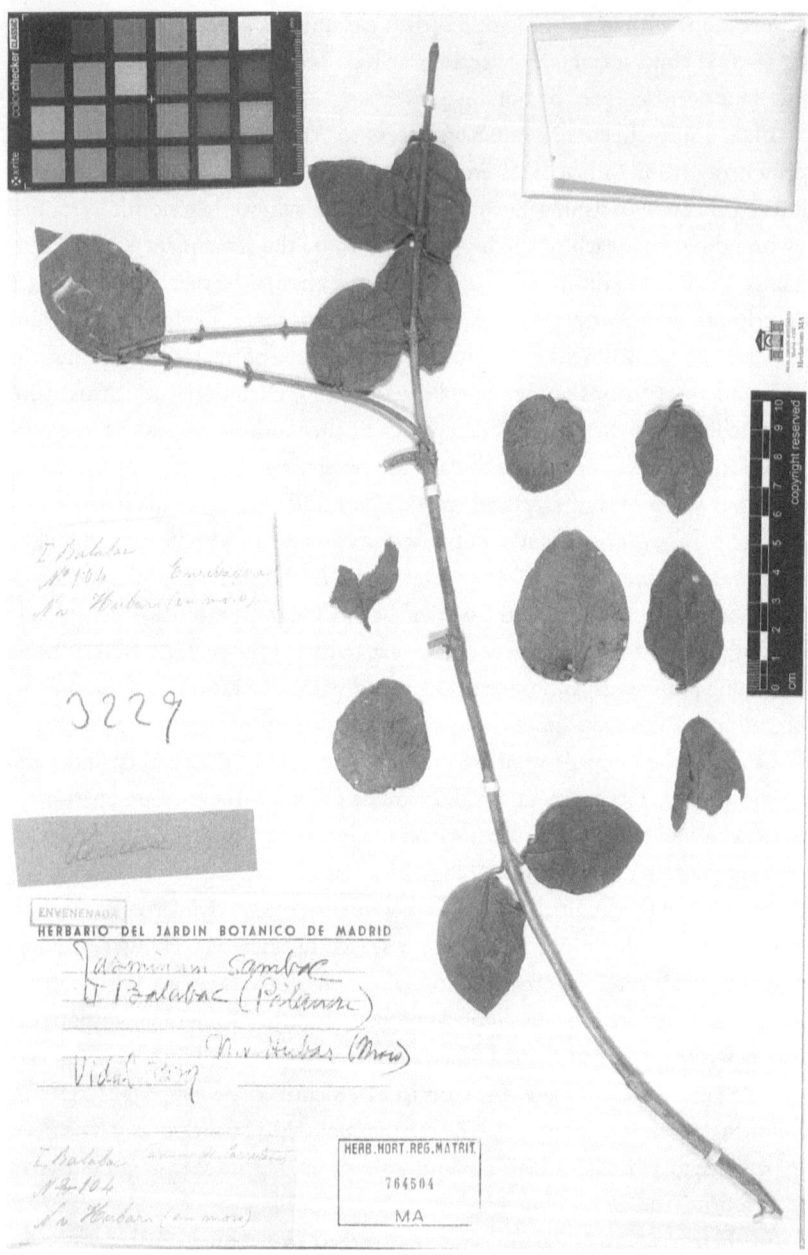

FIGURE 3.2. Herbarium sample of *J. sambac* from Sebastian Vidal y Soler's specimen collection retrieved on Balabac in Palawan. Notations indicate the name "hubar" ("in Moro") for the plant. Image and reproduction permission courtesy of the Herbario del Real Jardín Botánico.

a linguistic formation specific to the cosmopolitan city. "Manila" most likely referred to the language spoken solely in the colonial capital—a mixture of Tagalog, Spanish, and possibly Hokkien—and could have been any one of the languages of the marketplace.[43] That a unique plant name with such specificity may have developed in the capital's milieu gestures to the vibrancy of the human-plant engagements that raised the sampaguita to prominence and to the indelible impression the very name "sampaguita" would have on later generations.

I. Floral Invocations

Debuting a Particular Femininity

A survey of printed cultural production in the late nineteenth century demonstrates the frequency with which writers, composers, and visual artists invoked the sampaga and sampaguita. Often, the two defined—and were defined by—a prototypical femininity. Charming and quaint, young and alluring, ideal women and the flowers were one and the same.

The emotive, at times syrupy, stylings of such cultural production developed over the course of the century. Capital and enhanced access to education contributed to the literary eruption. Tagalog poets, in particular, wore a new cosmopolitan urbanity, which in previous generations had been reserved primarily for the peninsular and creole elite. Borrowing from the rhetorical flourishes of Spanish poetry, writers embraced the courtly romance tradition, marked in part by a "cloying lugubriousness in pining for impossible love." According to Bienvenido L. Lumbera, the folk talinghaga, or metaphor, gave earlier Tagalog poets before this time an analog drawn from everyday experience through which to convey the complexities of human existence. Thus, flowers, animals, and implements for weaving were incorporated into short poems to express more abstract states. By the nineteenth century, poets suppressed the talinghaga, "a unifier of sensations and ideas," in service to apostrophe, personification, and emotional extravagance found in Spanish writing.[44]

By the late nineteenth century, as Lumbera relates, propagandist poets, chiefly Marcelo H. del Pilar (1850–1896), who were "'schooled'" by the colony's finest institutions, aimed to "de-emphasize" formal learning in service to folk literary forms.[45] The cultural material I now discuss was largely produced during this period, as learned women and men straddled foreign genres within an atmosphere of political awakening. Across their sources, the sampaga and sampaguita appear as cultural anchors, mooring readers and listeners to the local environment. Their works resonate with the poems of the Ilokana writer

Leona Josefa Florentino (1849–1884), whose writing I also examine, and who was remembered for her acutely local creative sensibility. Still wound in ornate language, the popular species communicated the perils of love, the criteria of attraction, and the stages of womanhood.

Ahead of "Jocelynang Baliwag," the 1879 musical composition "La flor de Manila," by Dolores Paterno y Ignacio (1854–1881), circulated among an elite section of society.[46] The child of a wealthy Chinese mestizo family, she studied at the all-women's Real Colegio de Santa Isabel in Intramuros. Her "La flor de Manila" is a danza, or a genre of ballroom dance. Its accompanying lyrics operate as an ode to what the title denotes: "the flower of Manila," the sampaguita. The title immediately summons botanical imagery and geographical specificity, while its lyrics bring forth a refined woman. Paterno's brother, Pedro, author of the *Sampaguitas* (1880) poetry collection and the novel of Philippine customs, *Nínay* (1885), presumably penned the lyrics, which begin by presenting a comely flower:

> Sampaguita gentil que halagas
> Con tu aroma mi filipina
> Sampaguita flor peregrina
> ¡ay!, que en tus trenzas bordando estás
>
> Genteel sampaguita how you flatter
> with your aroma, my filipina,
> elegant flower sampaguita
> ah, in your tresses you adorn.[47]

At first listen, the lyrics praise the sampaguita's sensory qualities. Its perfume does nothing less than "flatter," while its very refinement has the capacity to "adorn" (or literally, "embroider") or make more beautiful. However, the original word choice also presents a lyrical playfulness, as double meanings render a collective understanding of femininity. The Spanish word "gentil" could mean handsome or pleasing, kind or courteous, but an older definition that has since fallen into disuse refers to one's belonging to clan, a people, or a nation—the opposite of an extranjero or foreigner.[48] This definition shifts the danza's overt meaning: the flower becomes part of a collective of some kind. The stanza carries on by crooning to the sampaguita as "my filipina," which likens the flower to a Filipina woman. The phrase, however, can also suggest that the flower is *of* the Philippines, thereby instilling the property of being "filipina" following the lexically feminine "la flor." Simultaneously then, the flower is both gendered and associated with the land. Moreover, not only is it *of* Manila. It is of the Philippines.

Another doubled meaning emerges with "peregrina." Its first signification comes through the word "elegant," as I have translated it.[49] A second arrives by way of its other translation: pilgrim. Intriguingly, even if the sampaguita is the flower of the capital, one imagined as growing in a particular place, it moves, specifically on the bodies of those who decorate themselves with it, as the remainder of the danza details. This "pilgrim" flower, dynamic because of its wearers' movements, travels and thereby captures more collective esteem. These lyrics permit several interpretations of the sampaguita, especially when one considers how later translations would reflect nationalist overtones in Tagalog and English.[50]

The Vigan-born poet Florentino likewise produced a number of works that culturally co-constituted the sampaga and the ideal feminine. She wrote in Ilocano and Spanish, and her writings were famously included in the *Encyclopédie internationale des oeuvres des femmes* (International encyclopedia of women's works) in 1889.[51] Her son, Isabelo de los Reyes, published her works in his *El Folk-lore filipino*, the two-volume compendium that sought to catalog archipelago-wide folklore. De los Reyes viewed folklore as an emergent field that could spur "self-critical understanding" built from an archive of local knowledge.[52] Florentino's poetic deployment of the sampaga—and the very inclusion of Florentino's poems in the elaborate publication—further imbued the sampaga with a singularly Philippine nature.

Florentino's "Coronación de una soltera en sus dias" (A maiden's crowning) regales a young woman on the occasion of her birthday while also issuing guidance on the preservation of purity and modesty. The poem likens the woman to the bud of a sampaga: after having grown under the loving nurturance of her parents, Mellang (a nickname for Emilia, the name of the celebrant) has now blossomed, with enviable beauty to behold and treasure.[53] As in the opening stanzas to "Jocelynang Baliwag," the subject of the poem is in her procreative prime. Here the flower concretizes this association. Unlike other structures of a plant—a stem, leaf, seed, or bud—the newly opened flower conveys natalist capacity. It signals a readiness for fertilization before producing a plant's fruit.

This "human/floral conflation" had not only dominated the science of Anglo-European botany but also literary portrayals of romance and courtship.[54] Anthropomorphic characterizations of the Philippine environment, however, did not germinate from European botanical romanticization alone. Male and female plant forms are believed to have been (and to be) prevalent among "unhispanized" polities.[55] Mount Makiling in the province of Laguna has been historically likened in form—from its curves to its cascades—to an ancestor spirit or diwata, Maria Makiling. The ilang-ilang is said to have sprouted from the grave of a lovesick girl, Ilang, whose spirit took the form of the fragrant flower. The talinghaga was not simply for "poetic ornamentation" but rather, as Lumbera explains, to convey "human experience [by] seeking out analogies in life and Nature."[56] By the late nineteenth and early twentieth centuries, Philippine writers waxed didactic, cautionary, and affectionate about the Philippine-born woman as a flower. In the essay "Las mujeres y las flores" (Women and flowers), de Los Reyes himself opines on their similarities. A scentless flower, for example, is a contemptible woman. Neither can prevent the scourge of age, but matrimony—like dewdrops—can revive any maiden.[57] His meditation on the value of a woman, parallel to that of a flower, is rooted in the reproductive capacity of both. Matrimony may save a woman beyond her prime, but the possibility of bearing fruit, human and vegetal, was the most prized feature of womanhood (or planthood).

Florentino reiterated much of the same, even provocatively differentiating the jazmín from the native sampaga. Florentino, like de los Reyes, allowed the imported word to stand independently from the Philippine one. Pedro Serrano Laktaw's (1853–1928) 1889 *Diccionario hispano-tagalog* (Spanish–Tagalog dictionary) reinforced this departure: the dictionary translates jazmín as simply "jazmín."[58] His translation suggests that the Tagalog language absorbed the foreign word for jasmine as a general category, distinct from the sampaga and sampaguita, which do not have a Spanish equivalent.[59] Florentino's choice to distinguish the two in "Coronación" could have been a poetic move to retain assonance: "rimmangpaya," "Ina," "curangna," and "sampaga." However, Florentino praised the sampaga in a breath of promise and beauty, whereas in "Felicitación satírica" (Satirical congratulation), she portrayed the withering jazmín as a running metaphor for a woman approaching spinsterhood at the age of 28, whom the poem advises to maintain a youthful air against the contempt of the inevitable. Conversely, her poem "Declaración simbólica" (Symbolic declaration) presents the sampaga as an adornment for sex appeal: it can be the quintessential young female or an attractant, one that incites "volcanic love" in men, as de los Reyes translated.[60] The jazmín, on the other hand, ominously symbolizes stages of womanhood.

Because Florentino was not formally educated, de los Reyes suggested her poetry was original to her, "not molded by European styles" but instead informed by the "muddled and unsightly" character of the Ilocano comedy libretti and thus reflecting a "genuinely Filipino style."[61] In *El Folk-lore filipino*, de los Reyes blithely remarked that in a compendium of native folklore, poems like hers are "worth little or nothing" because of their purely Filipino or Ilocano tastes, which may not align with European artistic inclinations.[62] Notably, de los Reyes's project was not crafted as an appendage to Iberian or European folkloristic studies. It was, as Mojares argues, a "resource for nation formation and not something merely ethnological" for foreign consumption.[63] Florentino's poetry was projected as part of a totalizing Filipino tradition, fundamental to reformist strategy for the Philippines' acceptance as a Spanish polity.[64] In this light, Florentino's poetry warrants two readings: first, for its gendered and gendering musings on the sampaga, and second, for its presentation in an entire folkloric lattice of the Philippines. In a compendium of this kind, that the flower could so generally refer to a woman—or that a woman could be so gratuitously compared to the flower—implies that the comparison is not only Ilocano but also Filipino.

A Plant So Generally Known

Depictions of the sampaga and the sampaguita would continue to communicate its universal popularity. In 1892, Trinidad Pardo de Tavera published *Plantas medicinales de Filipinas* (*Medicinal Plants of the Philippines*), a catalog of plants with allopathic and palliative virtues. Published in Madrid, the book provides botanical descriptions akin to the classifying work of the Comisión. Pardo presented data in a style common to colonial botany of the time: arranging plants by family and beginning each entry with a Latin name followed by common nomenclatural equivalents. Philippine elites like him were well aware of late nineteenth-century botany and its developments. Apprised of the Comisión's work, Pardo was a close colleague of García, who himself was highly regarded by José Rizal and journalist Graciano López y Jaena (1856–1896) following his rice exhibits in Europe.[65] In Pardo's writing, in particular, Latin binomials appear aside common plant names to present locally growing flora to audiences versed in international taxonomy. Following summaries of medicinal applications, morphological descriptions offer a rather technical view of plants and their constituent structures. Pardo described anonas, for example, as a fruit "without visible tubercles" and achuete as a tree with "corollas comprised of five pink petals" and "calyces of orbiculate sepals."[66]

Plantas medicinales may at first seem to conform to the expectations of botany, yet Pardo departed from these by considering the sampaga and the sampaguita the same species: the former from Tagalog, the latter from "Philippine Spanish." He provided no other local names. In place of a typical description, he wrote that the *J. sambac* was "perhaps the most esteemed flower of the Philippines ... for which description is useless because of its unmistakability among natives." He gave no morphological language, wasting no ink to define the plant's physical features. Because it "grows in all gardens," any native of the Philippines can determine the species with absolute certainty. After all, it "is so generally known."[67]

Pardo made no such suggestion for most other indigenously growing or even widely cultivated species, such as tobacco. Instead, an assumed collective familiarity surrounds the sampaguita. This familiarity gestures to anthropological research that has emphasized the significance of "covert categories" within folk plant taxonomies. In their study of Tzeltal communities, Brent Berlin, Dennis Breedlove, and Peter H. Raven emphasize the importance of "unnamed taxa"— that is, botanical categories that do not have conventional monolexemic labels. During their field interviews, Tzeltal informants grouped plants with species-specific names into otherwise unnamed categories. These groupings revealed hierarchical taxonomic structures unknowable to an observer simply focused on plant names.[68] For Pardo's *anti*-description of the sampaguita, the flower has a specific name that resonates with the collective imagination and, therefore, requires no actual description. The covertness at play then is not in the name itself but in the way the flower may be identified and described—within a sovereign vernacular uneasily accessed by the foreign observer. Contrary to imperial systematics, as Pardo's entry suggests, formalized description of the flower's form is unnecessary because other cues—visual, olfactory, or tactile—make the flower known covertly to the native viewer and perhaps, not so to others.

Pardo may have saved himself the trouble of enumerating the plant's structures by making such a bold statement, but establishing the sampaguita as something different is telling. He completed *Plantas medicinales*, a project commissioned by the state, over the course of two years while in Europe and crafted the work for a learned Hispanophone audience. Even so, he conveyed regard for Philippine-based herbalists, whose own process of obtaining knowledge of the therapeutic virtues of plants was not altogether dissimilar from scientists' drawing conclusions from observed phenomena. "Empiricism," he explained, gleaned from "daily experience" becomes knowledge transmitted over generations. Like de los Reyes, Pardo intensely labored to revise the existing corpus of information on the Philippines. Elevating native expertise could thus validate a domain of knowledge otherwise perceived with "contempt" by practicing physicians.[69] These assertions, therefore, cast his description of the sampaguita in another light. To him, there exists in the Philippines an "unmistakable" knowing, one derived from everyday encounters with the plant. Without a technical description, a foreign observer would fail to conceive of the plant's form. Instead, Pardo placed such knowing of it in the minds of Philippine-based herbalists and natives with whom a foreigner must consult to locate the therapeutic sampaguita.

In Pardo's estimation, foreigners had a flimsy grasp of the flora in the Philippines and the philological nuance surrounding them. Years before *Plantas medicinales*, he published *El sanscrito en la lengua tagalog* (Sanskrit in the Tagalog language) in 1887, diving into Tagalog philology, confirming and refuting previous works on the topic. On the sampaga, he corrected Dutch orientalist Johan Hendrik Caspar Kern's (1833–1917) 1880 list of Sanskrit-derived Tagalog words that suggested "sampaga" comes from "champaka," a term that references flowers of the *Michelia* genus or magnolias. Pardo denied this derivation, maintaining that Kern was unaware that the words sampa*c*a and sampa*g*a locally differ. Sampaca, he clarified, refers to a magnolia. Sampaga, "more commonly called sampaguita," is no such thing, *nor* is it "the jasmine mentioned in Tagalog dictionaries. It is a little flower with a very refined perfume and an extraordinary whiteness."[70]

These creative and intellectual assertions demonstrate how formally educated elites made authoritative statements against a legacy of foreign understandings of the Philippines and its environment. The sampaguita was central to these statements. The 1893 illustration "¡Sampaguita!" by Félix Martínez y Lorenzo (1859–ca. 1916) captures this point well (see figure 3.3). Martínez trained in fine art in Manila and on the peninsula and like his contemporaries, he contributed roughly fifty plates to the revised *Flora de Filipinas*.[71] He also was a principal artist for *La Ilustración Filipina*, a Hispanophone Manila serial, to which de los Reyes was a regular contributor.[72] In *La Ilustración* and elsewhere, Martínez garnered acclaim for his skill in the tipos del país style.[73] In Spain, the art form satisfied the tastes of wealthy urbanites seeking depictions of the "provincial other."[74] In the colony, an elite gaze angled onto depictions of what would have been considered common to the Philippine quotidian.

In 1893, *La Ilustración* published Martínez's "¡Sampaguita!" The image features a female vendor dressed in a simple baro't saya or blouse and skirt ensemble and standing barefoot upon grass. The vendor's hair is tied back loosely, with stray strands falling to her right ear. Beside the vendor is an outgrowth of live bamboo, and farther afield are a single pronounced palm tree and two elevated huts built from cut bamboo and palm fronds. As natural emblems of bucolic life surround the vendor, she looks directly at the viewer as she carries garlands of sampaguita strung and ready for sale.

Her rurality contrasts images of higher-class elite representations of women from the time, including the more famous portraits by Juan Luna such as "La Bulaqueña" (1895), which champions a certain modest femininity. Nonetheless, both images exude sensuality. In Martínez's depiction, the vendor's image is bound to her class as she clashes with the idea of a refined, demure woman. Her exposed shoulder, her gentle smile, and her offering of freshly plucked and strung

FIGURE 3.3. "¡Sampaguita!," by Félix Martínez, ca. 1893, as published in *La Ilustración Filipina*. Reprinted in Ramon Zaragoza, *La Ilustración Filipina, 1891–1894* (Manila: RAMAZA, 1992). According to Blas Sierra de la Calle, this sketch was the precursor to Félix Martínez's *Vendedora de sampaguitas* (Sampaguita vendor), a painting featured at the Exposición Regional de Filipinas (Regional Exposition of the Philippines) in 1895. See Sierra, "Félix Martínez," 62–63. Reproduction permission courtesy of Patricia Maria Araneta.

sampaguita all draw attention directly to her youth. She raises a garland in her left hand, and intriguingly, the detail of the vendor's fingers is much less pronounced than that of her right hand. The faintness of her fingers blends with the garland, making it less clear where her smallest finger ends and the flowers begin. The title alone then begs consideration. Is the sampaguita garland the eponym of the image? Is it the vendor? Or is it both? Given the cultural production coincident with Martínez's image, viewers may gather that a customary glimpse of the Philippines sublimates both woman and flower, the Filipina vendor and the everyday.

Taken together, the late nineteenth-century invocations of the sampaga and sampaguita distinguished both the colony and its scholarly and cultural production. These acts largely recited gendered and gendering portrayals common to what others have observed of patriotic rhetoric elsewhere.[75] As Raquel A. G. Reyes has shown, the ilustrados themselves grappled with the popular ideals of gender in elite Manila vis-à-vis notions of gender with which they were confronted in Europe. Although the ilustrados were on the whole proponents of women's social advancement, the "uninhibited 'Modern Woman'" was unsettling. For them, the feminine ideal to be mapped upon the Filipina woman was, at its core, conservative.[76] Elites co-constituted this ideal in my reading with a particular flower, which was further projected as an element of the common Philippine experience. These projections were, however, limited to the elites who produced them and to the small Spanish-reading audience in the Philippines (roughly 5 percent of the population) that could access them.[77]

For some, the flower would eventually be defamatory: Pedro Paterno's detractors cynically called him the "sampaguitero," a nickname gleaned from his *Sampaguitas* collection, after his arbitration of the Pact of Biak-na-Bato between the Spanish colonial administration and the revolutionary government in December 1897.[78] His negotiation of the peace agreement put an end to the first phase of the revolution against the Spanish. He allegedly requested that sampaga be scattered on his grave for his successful execution of the negotiations. Soon thereafter, however, he tried to ennoble himself within the Spanish colonial government, which attracted public ridicule.[79] Indeed, the flower would be invoked for still more purposes during the US colonial period.

* * *

In 1901, an English translation of Pardo's *Plantas medicinales* was released. By then, Pardo had begun to collaborate with the US colonial government after its 1898 acquisition of the islands. In the Anglophone version of Pardo's work, translator Jerome B. Thomas Jr. changed Pardo's original entry for *J. sambac* to include a conventional botanical description using morphological vocabulary. No longer reliant on a native informant, a reader could identify the plant by its "entire, glabrous leaves" and its "long, awl-shaped calyx." The entry also proclaims, "The flower is the most popular and beloved of any in the Philippines (and is commonly referred to as the *national* flower)."[80] Thomas's translation is immodest, as this phrasing does not appear in the original text. In the context of the earliest years of US colonization, that the Philippines could approach nationhood remained an unrealized aspiration—both for the insurgents and for the colonists who would impose their version of what exactly a Philippine nation would entail.

As the United States colonized the archipelago, violent subjugation was followed by sweeping civilizational tactics to prepare the colony for independence. Governing infrastructure, education, and civil society were molded to fit nationhood styled under US auspices. These measures coincided with the policy of "Filipinization," an initiative implemented under Governor-General Francis Burton Harrison (1873–1957), to prepare natives to fill the seats of the bureaucracy. During these decades, state-building projects steered by Filipino and US functionaries contrived a Filipino nationality based on volumes upon volumes of surveys to remake the archipelago an "object of knowledge and control." As state cultural agencies branded particular cultural forms as fundamentally Filipino, the revolutionary politics of the previous century transformed into watered-down histories in service to a brand of civic nationalism.[81] Though US colonists continued to encounter unexpected and unruly forms of patriotic expressions that challenged their own visions of nationhood, aspects of civic life would be molded to reiterate ideas of what exactly comprised the Philippines.[82] Of the early twentieth century, explains Mojares, "There was much enthusiasm for creating the substance and symbols of a canonical nationalism."[83]

National symbols arose, among which the sampaguita was essential. In the United States, the invention of state flowers emerged in the late nineteenth century. New York allegedly initiated the state flower movement but state adoption of floral emblems likely materialized around the planning and implementation of the 1893 Chicago World's Fair. The fair's "lady managers" advocated for floral emblems to represent each participating state, and Oklahoma was the first to legislatively approve a state flower.[84] In the Philippines, the elite production of the late nineteenth century had a staying role, even if transmuted to meet a new authorized form of US-like nationalism. Thomas's translation was only one of many publications that deemed the flower an especially significant emblem. An English grammar tried to teach the comparative degree with the following example sentences: "5. The sampaguita is *more beautiful* than the ilang-ilang. 6. The ilang-ilang is *less beautiful* than the sampaguita."[85] Poetry continued to tout the sampaguita as "widely prevalent."[86] In 1930, librarian Eulogio B. Rodriguez published *The Legend of the Sampaguita: The Filipino National Flower*.[87] (He followed with a book on Francisco Balagtas [1788–1862], the "national poet.") A year before the United States granted commonwealth status to the Philippines, Governor-General Frank Murphy (1890–1949) declared the sampaguita the national flower of the Philippines. The complexities of its stature as a culturally significant emblem of an elite section of nineteenth-century society were distilled into the present national icon.

Before its declaration as a national flower, the sampaguita (and the sampaga) existed as an object of scientific, intellectual, and artistic fascination. Foreign observations of the flower eventually led to its identification as the *J. sambac*, though linguistic forms rooted in the plant suggest intra-action well beyond what could be articulated through a Latin binomial. The binomial, however, helped Philippine-born intellectuals attach the flower to a species that could be identified more readily, though in some respects, would remain beyond sterile vocabulary. The late nineteenth century witnessed an influx of cultural production that upheld the flower and a certain brand of Filipina femininity. These were part of wider knowledge-projects to stress assertions that could counter a legacy of foreign representations of the archipelago.

As civilizational tactics unfurled under the United States in the early twentieth century, so did efforts to botanize the islands. Botany was one way that the United States could demonstrate its imperial weight in the Pacific as it took over the Spanish territory. While laboratory work and systematics were foundational to demonstrating intellectual aptitude, economic impulses propelled botany work. Botanical science for the United States, therefore, was a precursor to envisaging the Philippines as a commodity landscape.

4

Woven Transformations

MUSA TEXTILIS Née. Manila hemp or Abaká.
Musa textilis is probably the most important cultivated plant endemic in the Philippines. It produces the premier cordage fiber of the world.
—WILLIAM H. BROWN, "Philippine Fiber Plants," 1920

Around 1907, a weaver named Oleng parted with a section of fabric she had just completed on a backstrap loom. To produce its fine pattern, she dyed its abacá threads before weaving, having bunched and tied them with her fingers to prevent plant dyes of carob and auburn from smothering her motif. Usually, she would have dipped the woven fibers in ash, rinsed the fabric in a nearby river, and repeatedly ironed it with a mollusk shell, softening it to the touch. This time, however, her husband sold the piece (see figure 4.1) to a US anthropologist completing fieldwork in their locality who later recounted Oleng's hesitation.

FIGURE 4.1. Detail of fabrics obtained from Oleng, a Bagobo-identified weaver. A warp-faced cloth of natural, red, and brown ikat-dyed threads features interchanging motifs and patterns. To its side is another segment sold in the same transaction. "Ine and bata strips for tubular skirt," 70.1/5439 A + B, Laura Watson Benedict Collection, American Museum of Natural History. Photograph by author. Reproduction permission courtesy of the Division of Anthropology, American Museum of Natural History.

Among weavers, selling an incomplete textile was believed to be ill-advised. Therefore, she performed a rite over the piece with a single areca nut, a betel leaf, and a cup of water, transferring its spirit to a different but identical skirt. This act, she explained, would protect her from sickness as a consequence of the uncustomary transaction.[1]

Undoing the Botanist's Inventory

Weaving on the islands that would become the Philippines incorporated knowledge of plants for much of its history, with the earliest examples from the second millennium BCE.[2] Bamboo loom bars, wooden implements, rattan backstraps, bast fibers, and color-rich flora were common in the world of a textile weaver. A historically feminized domain of expertise, a weaver's plant taxonomy included raw materials, dye concoctions, finished tools, patterns, and fabrics at different stages of completion. Weaving was—and continues to be—embedded within ritual practice, trade, clothing, and custom. At the onset of colonial contact, Spanish and eventually US observers recognized the widespread nature of textile production and the commercial value of the very plants at weavers' regular disposal.

Following US takeover, the most active botanists sent to the newly acquired colony wrote on agriculture, manufacturing, and exportable plant goods. Akin to their Spanish predecessors, US botanists had a hand in economic instrumentalization of the Philippine environment. Pages and pages of their tracts focused on ventures associated with plants, muddling what has been a historiographical simplification of a singularly economically driven Spanish scientific operation vis-à-vis the purely "remarkable scientific progress" accomplished by the United States.[3] Botanists, usually writing in Manila and conducting fieldwork throughout the islands, folded fibers and dyestuffs into the larger fabric of exportable commoditization. Their published works read as indices, tables, and compilations of useful flora. In such inventories, as in this chapter's epigraph, a plant's Latin name usually appeared with its local equivalent(s) and a summary of uses, yet entries immeasurably distilled or overlooked the considerations a weaver like Oleng would have weighed when transforming abacá into a sellable good.

This chapter emerges from entangled knowing and making amid the imposition of colonial knowledge. I trace two parallel threads in the beginning of the twentieth century: US colonial botany's start in the Philippines and one anthropologist's fieldwork among a community of Bagobo, particularly weavers, in Mindanao's Davao Gulf. The weaving traditions and related material culture I incorporate in this chapter come from a group of Tagabawa-speaking Bagobo, one of three subgroups of Bagobo peoples residing in southern Davao. At the start of US Empire in the Philippines, anthropologists and ethnology enthusiasts arrived in Mindanao alongside botanists, military personnel, and bureaucrats. The Bagobo were one of the several animist or "non-Christian tribes," in the parlance of US colonists, who populated—and continue to populate—the island zone that had challenged complete Christian conversion and colonial-bureaucratic engulfment. Several foreign arrivals, regardless of disciplinary affiliation, collected weavers' objects. They purchased or traded for cloths, skeins, and women's, men's, and children's garments among the Bagobo in and around the coastal town of Santa Cruz and surrounding Mount Apo, the tallest peak in the archipelago, which holds cosmological importance for local peoples.[4]

The objects I examine and their annotations were eventually assembled in New York's American Museum of Natural History, one of the most extensive US repositories for early twentieth-century collections on the Bagobo. They serve as brief snapshots of female weavers, young and old, experiencing settler-colonial transformations.[5] This US-based collection was obtained and curated by an up-and-coming US anthropologist, Laura Watson Benedict (1861–1932), who arrived in Mindanao, intellectually moved by the live Bagobo exhibit at the Louisiana Purchase Exposition in St. Louis in 1904. A well-studied example of racial-capitalist exhibitionism, the Exposition was a point of reference for inquiry among those in the US and for those who would be inspired to journey overseas.[6] Benedict's extant collection is one of the more detailed corpora on the materiality of Bagobo textiles and their production. Her curation points to processes of transforming plant materials, alludes to struggles to obtain material culture from her informants, and associates particular weavers' names with specific garments—a rarity among colonial-era textile holdings from the Philippines.[7]

That US colonial intellectuals brought their disciplinary baggage and commitments to the archipelago is no surprise. Botanists and anthropologists stuck their claims to what they knew best of the Bagobo and other peoples. I examine their intellectual production in tandem, knowing these disciplines always already divvied up and consequently obscured "the teeming life between the worlds of natures and cultures."[8] My transdisciplinary engagement, as Karen Barad would suggest, therefore pronounces "the fact that boundary production between disciplines is itself a material-discursive practice."[9] Colonial disciplines disarticulated weavers' knowing and making for their own ends while encyclopedically cataloging such processes as a form of salvage anthropology.[10] Perhaps no compilation of words, as Marian Pastor Roces caveats, can fully capture the universe of a finished textile.[11] Even with Benedict's more copious notes, I do not intend to uncritically parrot her accounts. Like the botanists I cover, she stood to gain social and material capital from her work. Instead, I juxtapose these different sources—publications, annotations, and material culture—to reexamine colonial botany from the partial vantage of another sovereign vernacular. Doing so allows me to show botany's disciplinary making. That is, by analyzing the very material that ostensibly distinguished the colonial botanist from the colonial anthropologist; or the single-variable plant from the affectively dense textile sovereign to the orthodoxy of plant systematics.

The material culture representative of one ethnolinguistic community, I should caution, is not clear-cut. In the early twentieth century, ethnologists endeavored to define the societies of the archipelago, a fundamental objective of US anthropology at the time. Languages, mythologies, habits, and everyday items became the stuff of precise racial, ethnic, and tribal indexing. The acquisition of material objects was itself a racializing practice between the white consumer and the native societies whose goods they acquired.[12] In contrast, contemporary textile specialists of the Philippines warn against equating specific objects to a specific ethnolinguistic identity because doing so might fail to account for linguistic, geographic, and material complexities that make (and made) identities slippery.[13] Instead, I re-curate the objects as reflections of ongoing change—not simple stand-ins for what being Bagobo might mean. What I thus bring to light is how such objects point to the bodily skill and processual knowing necessary to manipulate flora for multiple purposes. More critically, the objects also provide a sense of concurrent transformations with heightened foreign intellectual and financial enterprise.

I begin by discussing US colonial botany's ascent in the Philippines under the backdrop of the United States' debut as an overseas imperial force. I then detail the colonial administration's focus on the Davao Gulf as a center for abacá, a member of the banana family, also widely referred to as hemp, which became one of the colony's chief exports in the early twentieth century. I turn to abacá in Bagobo weaving and lifeways before describing how abacá agriculture and waves of foreign developers impacted local Bagobo communities. I analyze select objects and think through them in the context of colonial encounters. These objects, largely derived from plants, wove together skill, meaning, and expertise. Examining Bagobo objects and their annotations demonstrates the interest of US colonial botanists in the marketability of weavers' materials. I argue the collected Bagobo material can surface transformative settler-colonial currents coursing through the gulf, demonstrating more clearly colonial botany's role in enabling such currents. Others have looked at the tremendous shifts in timber, agriculture, and plantation land development during the US colonial period.[14] I am concerned here with botany's use of weaving for commercially extractive pursuits.

I. US Botany in the Philippines

Imperial Ascent

The United States entered the global scene as an overseas imperial power at the end of the nineteenth century. Yet its advent as such arose from a much longer history of liberalism and continental expansion dating from the century prior. The country arose as an independent nation-state, having declared independence in 1776 with the American Revolutionary War ending in 1783. Leadership under George Washington (1732–1799) and Thomas Jefferson (1743–1826) was "very conspicuously imperial," as both sought to break from the European imperial ilk and obtain contiguous territories claimed by native peoples and the Spanish and French. Amid the liberal atmosphere of the Napoleonic Wars, the United

States nonetheless mirrored other European empires as it established conditions of "exception" with respect to its own imperial expansion. Policies of exception ensured that the laws in place to protect the "liberty and prosperity of property-owning farmers" did not have to be extended to native or enslaved peoples subject to expansionism or plantation production.[15] The determined acquisition of key ports in New Orleans, San Francisco, and Seattle over the succeeding decades also clinched US entrepreneurs' trade interests in order to extend commercial markets to Asia. Jockeying for overseas Spanish and British colonial lands dotted the nineteenth century, including unsuccessful attempts to plant a foothold in the Caribbean.[16]

The 1890s was a crucial decade in the United States' self-fashioning as an overseas empire. International and domestic developments hastened this self-fashioning. As an imperial juggernaut, Britain faced its own decline as a naval power, threatened mainly by the rise of imperial Germany. What had originally been British belligerence in order to keep US international expansion at bay turned into an "alliance imperialism"—a kind of Anglo-American solidarity—that permitted the United States' climb as a naval, trade, and diplomatic power. This Atlantic alliance could ensure a "global territorial status quo" while allowing the United States heightened access to and influence over Asian and Latin American affairs. Both powers backed an international order unique to the time which, as Courtney Johnson details, saw to the "development of a new system of international law to govern interstate relations via arbitration rather than armed conflict."[17] Domestically, the US industrial economy mushroomed following the Civil War (1861–1865). But economic volatility, which manifested in fiscal alarm and industrial slowdowns, motivated economists to consider expansionist opportunities in order to protect the country against what started to bubble up as class unrest. Widening access to foreign markets could, in their minds, offer a destination for the country's industrial overproduction while buttressing "the survival of liberal capitalism."[18]

The Cuban War of Independence (1895–1898) occasioned an opportunity for the United States to intervene in Cuban–Spanish affairs. Sensational journalism of the time roused public dread over the nearby threat of Spanish aggression against the United States. Mounting concerns reached a high point with the explosion of the USS *Maine* in February 1898, kicking off the Spanish–American War fought that same year. Combat extended across the Pacific to the Philippines. The United States allied with local revolutionaries aiming to overthrow the Spanish stronghold and with Spanish surrender in sight, revolutionaries declared the archipelago's independence in June 1898.

For the Philippines, its revolution of 1896 and the Spanish–American War originally signaled the end of imperial governance and the start of local rule. Hard-fought battles across the archipelago—waged by forces of different forms and ideological shades, with both congruous and contradictory aims—had momentarily relieved the emerging nation of its colonial yoke. Not long after, convenors of the Malolos Constitution founded the First Philippine Republic. However, the US fleet soon descended upon the Philippines to forcibly occupy the territory. Beginning in 1899, the Philippine–American War brutalized the archipelago with combat, disease, and famine, resulting in the death of more than 220,000 Philippine civilians and fighters. The US military relied on tactics from the Indian Wars fought in the decades prior, applying the lessons of its continental expansionism to new tropical terrain.[19] Fighting was tireless, factionalized, and complex: some local forces acted against each other and collaborated with US military campaigns. While major forces against the US surrendered in 1902, contingents of Moro fighters in Mindanao continued to engage in skirmishes well after armed units and elites declared concessions in Manila.

War and Botanical Maneuvering

At the height of military encounters, US personnel wasted no time in beginning to evaluate the environmental, economic, demographic, and political conditions of the Philippine territory. Compared to its contemporaries and the former Spanish Empire, the US sought to represent itself as a more intellectually sophisticated, science-forward power.[20] This aspiration was especially visible in the country's botany endeavors.

In the final decades of the nineteenth century, a "new botany" had materialized in the US that departed from pure botanical description and plant collecting. Prioritizing plant morphology, ecology, and evolutionary systematics enabled practitioners to characterize the discipline as a more experimental, professionalized science.[21] This shift coincided with a "botanical Monroe Doctrine" that, like the foreign policy of the early nineteenth century, challenged European supremacy. Specifically focused on British and German rivals, US botanists sought to make the United States the leader of botany in the Western hemisphere.[22] The crop of botany personnel sent to assess the flora of the archipelago were an outgrowth of this project.

Among US initiatives on education, defense, and governance, the colonial administration backed the establishment of the Bureau of Government Laboratories, later renamed the Bureau of Science, in 1901.[23] One of the bureau's principal architects, Dean C. Worcester (1866–1924), envisioned a wide-ranging laboratory that could promote research and collaboration across scientific disciplines.[24] By 1904, the bureau's first director, Paul C. Freer (1861–1912), boasted the completion of the bureau's principal building between Calle Herran and Calle Padre Faura in Manila.[25] He celebrated how an individual scientist could find "himself in a scientific atmosphere and in contact with students of all branches, giving him a broader and more satisfactory career and bringing the government better results."[26] The bureau housed foresters, agriculture specialists, entomologists, botanists, pathologists, and chemists. Laboratory work for all the disciplines was preeminent, as was the descriptive and taxonomic work of its botanists, who initiated the bureau's herbarium collection.[27] Study of plants was foundational to Freer, such that he opined, "As the people depend upon products of the field and of the forests for so large a proportion of their sustenance and barter, a knowledge of the flora of the Tropics is essential, from both a scientific and a material standpoint."[28]

The Stuff of the Index

Deployed to the Philippines in 1902, US botanist Elmer D. Merrill (1876–1956) asserted, much like Freer, that "accurate knowledge of flora" was the "first essential for future successful agricultural and forestry work." According to Merrill and his colleagues, the "secondary" objective to study the economic and agricultural capacities of colonial flora relied on the "primary" work of classification, which would ensure that the United States could approach comparable levels of success seen in nearby colonial holdings, as the British and Dutch had demonstrated in Singapore and Java.[29]

Over the next two decades, botanists in the colony authored and co-authored with agriculturists and foresters troves of material on plants with economic value. In addition to much-coveted timber, they projected the economic importance of other "minor forest products," including those derived from palms, bamboos, and mangroves, among others.[30] Colonial scientists even prospected the heightening paper pulp industry in the Philippines, where, in their words, "raw material of good quality and at cheap prices is available." They considered the increase in print media in the early twentieth century—consisting of eleven dailies and eight weekly papers plus other periodicals in Manila alone—as contributing to domestic market needs and motivating the discovery of an alternative to the industry's import-dependence. In effect, excess cotton and other "waste" material from fiber-yielding plants such as abacá, sisal, and maguey all portended profit.[31]

A standard colonial-era compilation on useful plants reveals how Merrill and his colleague, William Henry Brown (1884–1939), whose writing on *Musa textilis* opens this chapter, assessed lokdó, or what they identified as *Dryopteris pteroides*, of Samar. To the botanists, the plant had "cord-like vascular bundles" used for basket weaving and fibers of "inferior quality."[32] Whether "inferior" refers to lokdó's tensile strength, wear over time, or utility compared to other bast fibers is unclear, just as the lokdó's broader significance is irrecoverable from the entry alone. This system of arrangement—separating constituent knowledge from its contexts—is rife in studies of weaving and dyeing in the Philippines. In the interest of specificity, tables and indices persisted in isolating plants from weavers' and dyers' emergent cultural, economic, and environmental considerations.[33] Each table or index was a tool of efficiency, including Latin nomenclature to facilitate foreign readers' comprehension. From one local knowledge system to another, as seen in Spanish-era publications, Latin mediated translation, further dictating each entry as the first listed name and subsuming local equivalents.[34]

Within the longer history of imperial botany, Merrill and Brown's style of cataloging useful plants was not especially new. Capitalist pursuits and botany had overlapped since the early modern history of the science. Species identification and evaluations of plants' value met the interests of the eighteenth-century English Empire, for instance, as it invested in botany as an "economic science" to exploit plants' agricultural potential.[35] US colonial botanists' work drew on a lineage of European imperial naturalists, who hoped to acclimatize introduced species for wide-scale cultivation, identify competitive alternatives to globally traded spices, and pirate plants like the *Opuntia* genus of cactus known to host the lucrative carmine-bearing cochineal insect.[36] In short, botanists' "peculiar grid of reason over nature," as Londa Schiebinger has aptly described, was lucrative.[37]

In particular, commercial magnates recommended growing and harvesting abacá in a "scientific manner" to ensure maximum productivity. Refined agricultural methods, knowledge of abacá varieties, pest abatement, and keen knowledge of the plant's fifteen-year lifespan and fiber-yielding pattern ensured that magnates could control production levels in spite of fluctuations in global market demand.[38] Circulars produced by the colonial Bureau of Agriculture (see figure 4.2) and periodicals regularly offered experiment-derived instructions while remarking on the untapped potential of agricultural lands. In Negros Oriental, US observers balked at families' relying on their land holdings "as one would draw money from the bank. If funds are needed the nearest and most convenient plants are cut out.... The result is they get less than one half the price they could get with a little more care and foresight." The province's abacá district, in the observer's estimation, would yield fortunes for "men with capital and brains."[39] Likewise, prospectors in the emerging timber industry perceived native forest dwellers as "less enthusiastic capitalists."[40] Seizing upon such opportunity, foreign enterprising initiative continued to make its way

FIGURE 4.2. A once known man standing in an abacá grove. The image appeared in various US-produced circulars. Image and reproduction permission courtesy of the LuEsther T. Mertz Library of the New York Botanical Garden.

southward to Davao where by 1910, "more than one hundred Americans [were] hewing their way into the heart of the never-trodden forests."[41]

Southward Stratagem to Mindanao

As botany personnel rifled for plants, Mindanao was expected to be "the richest of all in the Archipelago in the number of species and in the luxuriance of its vegetation."[42] Historically, Spanish-era naturalists and botanists had conducted more intensive studies on Luzon and in a few notable localities in the Visayas. Only by the late nineteenth century had Mindanao become a more surveyed region following the Act of Incorporation in 1851 through which the Spanish Crown had sought to quell the Moro south and widespread piracy in the area.[43] Spain, like other European empires newly armed with steamship technology, deployed vessels to expand its governmental reach in administratively remote regions.[44] Sebastián Vidal's work and that of the Inspección General de Montes reflect this southward stratagem. At the start of the US regime, the collected plant material from Mindanao, though meager relative to that from Luzon, suggested the environment's distinctive yet understudied qualities.

In 1903, the colonial government established Moro Province, a large territory that extended across Mindanao from the Sulu archipelago, through Zamboanga, across Cotabato, and up to Davao's eastern coast. Members of the military restructured the province, which had been ancestral homelands to Muslim sultanates and highland and coastal animist communities, including the Bagobo, into districts that were further divided into municipalities and "tribal wards," using methods of land occupation and native subjugation akin to those in the US continental frontier.[45] Agricultural development, as Alyssa Paredes narrates, in cattle, rubber, pineapple, and as I discuss later, abacá, "served the cross-pollinating purposes of American racial imperial and Filipino settler colonial projects" in the region, thereby incorporating local communities into "new ethnic hierarchies and systems of agrarian productivity."[46] An effort to modernize non-Christian peoples, the formation of Moro Province incorporated the zone into the proto-national Philippine fold.[47] Until then, only a smattering of missionaries, naturalists, land surveyors, and colonial functionaries had lived in what would become Moro Province's Davao District. The district and its coastal town of Santa Cruz, founded by the Spanish in 1884, became a way station for new arrivals to the region in the early twentieth century.

In this severely militarized moment, at least 5,000 people from the United States descended upon Moro Province.[48] Ongoing armed conflict, rather than impeding the arrival of plant collectors, botanists, and botanical enthusiasts, propelled it. In 1903, the New York Botanical Garden (founded in 1891) sent Robert Statham Williams (1859–1945) to amass specimens in Santa Cruz. Interested in edible and useful fungi, among other species, systematist Edwin B. Copeland (1873–1964) collected in Catalonan, Malalag, and Mount Apo, and along the Sibulan River in 1904. Copeland's collections complemented the work of his colleagues, such as US independent collector Adolph D. E. Elmer (1870–1942) and German botanist Otto Warburg (1859–1938), who also participated in the collecting clamor in the south. Within a little over a decade, Mount Apo was one of the most botanically explored regions in the Philippines, comparable to the area surrounding the College of Agriculture (founded 1909) in Laguna and Mount Mariveles in Bataan, both relatively more accessible to those based in Manila.[49]

Intensified foreign presence also included US bureaucrats and planters—sometimes one and the same—looking to capitalize on Davao's soil, which enjoyed relatively consistent rainfall throughout the year. The US deployed Edward C. Bolton (d. 1906), a lieutenant in the US Seventeenth Infantry, as the district's first governor-general. After a plant-surveying expedition to the district, Copeland named a species of fungus, *Agaricus boltoni*, to honor the appointed official.[50] Bolton's administrative tasks included implementing the tribal ward system. In Davao, the system, in part, enticed upland peoples to labor on plantations on the coast. It facilitated the resettlement of peoples to villages that centralized labor supply, keeping laborers and their families near plantation sites.[51] Planters could then exchange goods, housing, and protection from rival tribes for labor.[52] Bolton selected a European planter to be "headman" of the Tagakaolo tribal ward, which encompassed other communities, including the Kulaman, Bilaan (presently transliterated as B'laan), and Kalagan. Residents' reactions to the new system of social organization that accompanied the development of plantation agriculture varied, but local opposition boiled as reports of planters' maltreatment of laborers reached US officials. A millenarian movement sprang as a pan-tribal effort to purge US presence in the region. Hostilities came to a head with news that Mungalayan, deputy headman of the Tagakaolo tribal ward, orchestrated the murder of Bolton and a discharged soldier-turned-plantation foreman, Benjamin Christian (d. 1906).[53] The sensational event prompted speculation among US settlements in and surrounding Davao that antagonism was building against the foreign presence, especially tied to the expansion of foreign-owned plantation lots.[54]

Transformations

Descending Anthropologists

Botanical and agricultural pursuits paired with ethnological efforts in the region. Mount Apo and its surrounding locales had motivated adventure and scientific surveying, paving a beaten path for other foreign arrivals. Fascination with Mount Apo likely dictated why most Spanish and US colonial-era documentation of the Bagobo focused on Tagabawa Bagobo communities. A common early route up the peak went through Tagabawa Bagobo lands.[55] Vidal wrote of the "great volcano" that towered over Davao's luxuriant countryside and took note of the Bagobo residing in the vicinity.[56] Bagobo also opened their homes over several days to Regino García as he studied Mount Apo's vegetation.[57] Early species records suggest that members of the local Bagobo communities encountered US collectors and provided their own plant nomenclature, and such ongoing contact facilitated wider ethnological work eventually conducted in the gulf.

One of the most studied ethnological pursuits during the US colonial period is the Louisiana Purchase Exposition, or the World's Fair, of 1904. The US government and private patrons envisioned the event to capture the expansiveness of the United States' contiguous and overseas empire. The Philippine Exposition Board, responsible for the material culture and live human exhibits that were to be featured at the fair, commissioned collectors to fleece the islands for objects and goods. In Davao, this project included two Philippine Bureau of Education teachers tasked with sending plants, samples of timber, fibers, fruits, and seeds to St. Louis.[58] At the Exposition's so-called "human zoo," one thousand people from the archipelago were present among other indigenous peoples of the US and those of other territories. The widely broadcasted event attracted nineteen million visitors during its seven-month run. Igorot, Visayan, and Bagobo peoples journeyed with US officials to Missouri to compose the "Philippine Reservation," in exchange, at least in the documented case of the Igorot, for remuneration of their time.[59]

Benedict, then a graduate student in an anthropology course at the University of Chicago, marveled at the thirty-eight-person model Bagobo village.[60] Particularly enchanted by Bagobo material culture, she trained in ethnological and museum collecting and display before embarking on a fourteen-month trip to Davao, combining her own funds with the support of a teaching post under the US War Department.[61] She was not the only collector who journeyed to southern Mindanao. The collecting fervor for native goods had already been a noticeable practice in the United States among private merchants and museum curators. The rise of Boasian anthropology, which advocated cultural relativism as opposed to social evolutionism, further sparked collecting practices that championed field research and could better capture the lifeways of specific groups as opposed to their teleological relationships to others.[62] During Benedict's fieldwork, she obtained enough fluency in the Tagabawa Bagobo language to conduct interviews with informants in Santa Cruz and around the south and southeastern slopes of Mount Apo.[63]

Abacá Landscapes

Benedict's fieldwork included recording oral accounts of how hemp culture informed Bagobo origin tales. Abacá weaving marked a cultural awakening, according to her informants, who recounted how their impoverished ancestors lived without the artistic refinement of patterned threads. Instead of woven garments, people wore sheets of bark from the trunks of coconut trees. Once equipped with the skill to weave hemp, Bagobo female ancestors dyed abacá fibers of clarets and sorrels. As an interviewer, Benedict observed a "vital relation

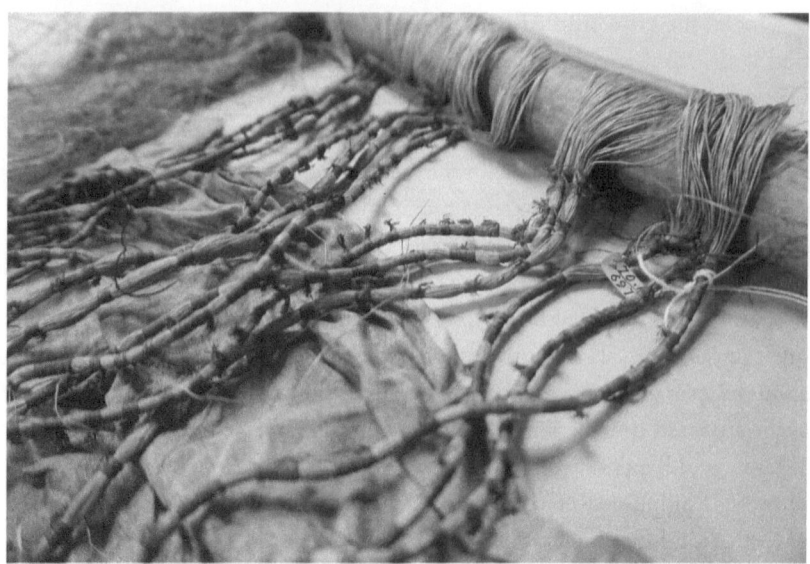

FIGURE 4.3. Detail of frame with abacá warp threads bunched and tied for dyeing, 70.1/6973, Laura Watson Benedict Collection, Division of Anthropology, American Museum of Natural History. Photograph by author. Reproduction permission courtesy of the Division of Anthropology, American Museum of Natural History.

between beauty in personal adornment and a virtuous character," which she suspected was rooted in the "golden age" stories of the "prehistoric Bagobo."[64]

In the early twentieth century, the fiber's versatility enabled Bagobo to convert it into toys, masks, baskets, garments, and dishware for ceremonial purposes.[65] Weavers obtained thread by scraping the plants' petioles to separate its individual fibers, hanging them to dry and bleach in the sun. They tied each thread to produce long, unbroken chains, ensuring that each tiny knot would not snap or come undone. Any of the ensuing processes could threaten the chain: threads were tied for dyeing (see figure 4.3), held taut between a backstrap loom bar and a weaver's body, and even beaten by a heavy section of polished wood so weft threads firmly interlocked with warp. According to weavers who spoke with Benedict, "pattern units" were distinguished by the number of clustered threads, such as "tolupolu" for thirty threads, "kapayan" for forty, and so on.[66]

Beyond the Bagobo, abacá factored into the weaving and sartorial practices of other communities.[67] Prior to European contact, Chinese traders long took commercial interest in the fiber.[68] In 1801, Spanish naturalist Luis Née (1734–1807), who had participated in the five-year Malaspina Expedition, dubbed what he observed to be the locally versatile species *Musa textilis*.[69] He encountered the plant in cultivated plots surrounding Mayon and Isarog volcanos and throughout Camarines, Albay, and Laguna provinces, though abacá had also been known to grow wildly in Mindanao.[70] Spanish trade contributed to the fiber's increased popularity. Given the number of endemic varieties in the Philippines, commercial plots throughout the second half of the nineteenth century featured abacá as a staple crop. Highland communities in Mindanao were also recorded to have cultivated and sold the crop to coastal merchants.[71] Its ascension as a global, trade-generating good intensified during the US colonial period.[72]

In 1903, colonial officials declared abacá the "most important industry" that "fear[ed] no rival."[73] The fiber's tensile and saltwater-resistant qualities made it especially appealing for cordage. A year prior, the US Bureau of Insular Affairs reported that the archipelago shipped 100,000 imperial tons of the fiber annually.[74] By 1917, the figure increased to 166,586.[75] Brown concluded that most of the other plant-derived fibers in the Philippines were important but would prove unprofitable as abacá outpaced the market performance of other bast fibers such as pineapple, cotton, bejuco, buri, nipa, and pandan.[76] Big and small landholders capitalized on Bicol and Davao as primary abacá plantation sites because the sloped landscape in the regions kept the rain-fed earth moist but not waterlogged—ideal growing conditions.[77] Larger land-owning incentives also stimulated hemp production to grow at increasingly competitive scales.

In search of the green gold, US, Japanese, Okinawan, Ilocano, and Visayan planters traveled to abacá regions in the south. Settlers from the Visayas, and to a lesser degree, from parts of Luzon, invested capital, oversaw business and legal transactions, and labored in fields.[78] Japanese planters arrived in Davao in massive numbers in the 1920s through the 1930s, working as estate holders, middlemen, and laborers.[79] Compared to other locations, Japanese-operated abacá estates developed in a competitive fashion, with increased mechanization to improve efficiency. Because many migrant workers, especially from the Visayas, were unfamiliar with abacá production, plantation holders could quickly acquaint them with their own practices. In contrast, the local labor force in Bicol had more extensive abacá cultivation experience and were less inclined to learn new techniques.[80]

In southern Davao, planters and working parties slowly displaced local Bagobo communities. Japanese settlers felled trees with both material and spiritual import to Bagobo communities on domains the government had called "public" but were otherwise under the stewardship of nearby Bagobo. As abacá plantations proliferated, smallpox and influenza epidemics killed many local residents from 1917 through 1918. Murders of Japanese settlers were not uncommon, and Bagobo-Japanese conflict erupted in spurts. During Benedict's fieldwork, colonial directives resettled Bagobo toward the coast.[81] Furthermore, the imposition of tribal wards enforced a monetary tax that required residents to find wage-based employment.[82] Many Bagobo began to participate in the cash-based economy and worked on plantations themselves, which were known for their relatively better compensation.[83] Some were observed converting their own plots to cultivate the banana fiber more intensively.[84]

Flexible Practices

In her assessment of early modern Europe, Pamela H. Smith characterizes craft or artisanal knowledge as a series of processes that are empirical, collaborative, produced within a community of experts, and responsive to the external forces of one's environment. The final characteristic, which underscores artisans' flexibility to variables such as weather or materials, also applies, I contend, to the colonial context of the Benedict collection. Thinking capaciously about the flexible responses in the objects and their accompanying annotations heeds Smith's call for serious consideration of the contingencies of "making knowing" and the "dynamic and emergent" quality of experiential knowledge. This approach is germane not only to historiography of the Scientific Revolution, to which Smith attends, but also to the history of colonial science more broadly.[85]

In the process of making, a craftsperson is "confronted with the grounding conditions and the enabling operations that go into making art but are usually not acknowledged in interpretation or attribution." These conditions and operations—from the materials, their qualities, and their manipulations— might not be as readily visible in a finished product.[86] For Bagobo textiles, the abacá pulp separated from fibers, the pulverized plant remnants stripped of coloring compound, the bits of thread fallen to the ground when finalizing the smooth edges of a piece, or the days upon days spent soaking a fabric in rich red ochre liquid are intrinsic to understanding making and knowing. Moreover, other facets of knowing come into view from the colonial circumstances surrounding practicing Bagobo weavers and the very collection of which their work may be a part. One can thus analyze objects as Bagobo weavers' flexible responses to their environments as both material considerations they made and as reflective of wider "geo-eco-political events" within the object itself.[87]

Flexible responses appeared, for instance, in Bagobo weavers' finished skirts, such as the magisterial three-panel ginayan or panapisan, as Benedict cataloged them. The finished skirt joins three woven sections at their selvages: a central piece (ine, or "mother") and two identical pieces (bata, or "children") that flank it. Intricately dyed with a host of detailed motifs, the finest ginayán were donned, displayed, traded, and cherished.[88] The ine panel's motifs differed

FIGURE 4.4. Detail of tapang sewn onto the ine panel. "Skirt, woman's panapisan," 70.1/5366, Laura Watson Benedict Collection, American Museum of Natural History. Photograph by author. Reproduction permission courtesy of the Division of Anthropology, American Museum of Natural History.

from the two identical bata: some of the motifs, as certain weavers pointed out to Benedict, were faunal and floral forms. The bata panels included intricate, narrow repeating patterns to complement the ine's larger motifs.

As Benedict observed, Bagobo weavers sewed a different patch of woven material into the ine, referred to as a tapang, that functioned as a protective component of the skirt to keep its wearer from becoming ill (see figures 4.4 and 4.5). Not all ginayan obtained by Benedict featured a tapang. In Benedict's estimation, the tapang's absence seemed to demonstrate a "gradual breaking down of an old custom in obedience to artistic instincts for symmetry."[89] The ginayan of figure 4.4 is a warp-facing plain weave. All three panels are sewn together, forming a tubular skirt. Threads of browns, creams, and reds flank the mother panel of predominantly burgundy and natural threads. The mother panel is noticeably shorter than the child panels, with a section of fabric added to it, which resembles the bata pattern.

Studying the Benedict collection, Judith F. Sibayan suggests that within the "spiritual environment of the Bagobo," objects maintained "their own individual potency." Textiles in particular acted upon weavers' and wearers' bodies, depending on the rites they followed or disregarded. Women wove among the

FIGURE 4.5. Detail of tapang sewn onto the ine panel. "Skirt, woman's panapisan," 70.1/5366, Laura Watson Benedict Collection, American Museum of Natural History. Photograph by author. Reproduction permission courtesy of the Division of Anthropology, American Museum of Natural History.

Bagobo, but not all women practiced weaving or could access complex dyeing techniques.[90] Certain patterns, for instance, were reserved for the bodies of elders and master weavers, with foreboding consequences for any younger Bagobo who broke convention. Adorned with appliquéd and embroidered designs, certain Bagobo were "thus cloaked with a whole arsenal of pharmaceutical, sacrificial, and magical agency."[91] With respect to the tapang, Sibayan infers that its inconsistency across the selection of textiles points to shifting tastes in fashion and ritual practice. Filomeno V. Aguilar has noted how the sacred or magical quality of particular cultural processes across the archipelago, such as rice agriculture, declined as commoditization increased.[92] In this case, one might further presume that demographic and economic shifts also contributed to the tapang's gradual disappearance, despite its significance as a protection against disease.

Census data from Mindanao in the early twentieth century can be spotty, particularly about inland animist communities. At the end of Spanish colonialism, however, an estimated 50 percent of Mindanao's total population consisted of individuals from other regions of the Philippines with the remaining 50 composed of Muslim (roughly 30 percent) and animist (estimated 20 percent) inhabitants. This domestic influx of settlers from the northern and central regions of the Philippines tacked on to Spanish efforts to quell Muslim dominance in the south. In addition to the US settlers who arrived during the military takeover of Mindanao, the Public Land Act of 1903 further encouraged homesteading, granting forty acres to individuals and upwards of 2,530 acres to corporations. Following 1903 and for the next three decades, Mindanao's population growth was above the colony's average.[93] In the first fifteen years alone, Patricia Irene Dacudao reports, the population rose from 65,000 majority indigenous peoples to 102,221, with "near equal" numbers of Christians and non-Christians.[94] Net migration pushed up these figures, and the principal area in Mindanao's south seen to gain substantial population was Davao. In a thirty-year period, Davao and the northern provinces of Mindanao—which border the Visayas region and therefore experienced a larger wave of migrants because of proximity—saw 1.4 million migrants. Benedict collected in the first three years of this demographic shift, and her Bagobo informants remarked on the changing dynamics in the community.[95]

Benedict herself associated the tapang's disappearance with aesthetic symmetry in the textiles that seemed to be more prevalent in the dress of nearby and newly arrived communities. She extended this observation to certain patterns, such as the plaids of the nearby Moro communities that were "finding favor with the Bagobo" in order to "please Americans."[96] Her annotations do not expand on this topic, but images from the US colonial period point to sartorial choices among elder Bagobo men, for instance, to wear US-style hats and collared button-down shirts or to put on military goggles while posing for photographs.[97] The novelty of such items was not lost on the Bagobo, who could demonstrate dynamic tastes amid the influx of new peoples, modes of dress, and what those modes implied. Social hierarchies that became apparent with the expansion of plantation agriculture may have thrown such implications into clear view. Bagobo men and women may have been motivated to assimilate to the fashion practices of other communities because of planters' variable treatment, social connotations, and aesthetic predilections. During her fieldwork, Benedict herself more generally concluded that the death of elder leaders and "the transference of entire mountain groups to provide native labor for American plantations" were "unquestionably" modifying Bagobo ways.[98]

At the same time, Bagobo weavers also made do with the inflow of different materials to incorporate in their works. As much as they had relied historically on abacá to yield finished textiles, other types of fibers proved alluring. Cotton in particular proliferated as a result of enhanced trade in town centers, and weavers in the region could readily obtain it. The fiber had a long precolonial history in native weaving traditions in the northern Philippines. By the early

FIGURE 4.6. Detail of "Man's hemp jacket, umpac ca mama," ca. 1900, 70.1/5281, Laura Watson Benedict Collection, Anthropology Division, American Museum of Natural History. A weaver designed beadwork over bands of cotton appliqué and abacá fabric. Tassels of colored cotton threads and "simarun" seeds, according to Benedict's notes, hang from the collar. Photograph by author. Reproduction permission courtesy of the Division of Anthropology, American Museum of Natural History.

twentieth century, government-run exchanges and plantation stores sold or traded basic goods and even traditional ware to particular indigenous groups. Trading posts cropped up, mirroring the civilizational tactics deployed by the United States on tribal lands back on the continent. According to one anthropologist, Bagobo communities produced extra abacá and brass goods to trade for cotton cloth and sartorial adornments such as beads and shell discs. Slowly, retail shops came to expect sales on cash or credit, distinct from the barter practices seen among these communities. Within a matter of decades, with the influx of cash and foreign proprietors, the "commodity-to-cash crossover" more fully materialized. Coincident with this were changes in material lifeways that, as Dacudao concludes, "compelled the tribes to join their lowland counterparts under the umbrella of consumerism in the guise of a decidedly American modernity."[99]

Plain and dyed cotton figured into weavers' finished abacá textiles, sometimes as supplementary weft, as warp threads, or as embellishments in the form of fabric bits or pompoms hanging from decorative tassels on blouses (see figure 4.6).

FIGURE 4.7. Detail of "Shirtwaist, old man's or old woman's," 70.1/5258, Laura Watson Benedict Collection, Division of Anthropology, American Museum of Natural History. Photograph by author. Reproduction permission courtesy of the Division of Anthropology, American Museum of Natural History.

Weavers designed fabrics that embraced the fibers' special composition and textures. Combining the two into a smooth cloth, or even as an appliqué, required knowing how the two fibers would interact with each other, and how to pull the abacá just taut enough for cotton twists to evenly blend with it.[100] As Benedict detailed in her notes, nearby Bilaan weavers embroidered with a mix of cotton and abacá, yielding a smooth finish with elaborate designs that Bagobo community members also wore (see figure 4.7). Other weavers used cotton material to weave full pieces. Enough of it had flooded the market in the first decade of the twentieth century that some communities could make toys from excess threads, such as a children's cotton spool to imitate the daily work of an elder weaver (see figure 4.8). As Benedict recounted, Visayan presence in the region also impacted the fashion sense of younger women, who started to prefer clothing derived from other materials than hemp.[101]

Synthetically dyed materials also gradually became a feature of Bagobo garments and wear, though the transition is trickier to trace within the transformative currents of the early twentieth century. The first synthetic dye was fabricated in 1856 in Britain, sparking the demise of the global natural dye market.[102] Synthetic dyes became a growing feature in domestic and international textile

FIGURE 4.8. Detail of "Spool with cotton," ca. 1900, 70.1/7168, Laura Watson Benedict Collection, Anthropology Division, American Museum of Natural History. Faint shades of green remain toward the center of the spool. Yellow and red threads appear on the right edge of the loop. Photograph by author. Reproduction permission courtesy of the Division of Anthropology, American Museum of Natural History.

production in the Philippines, and Bagobo, too, included synthetically dyed cottons in their finished products. A less time-intensive alternative to leaching color from plant matter, these dyes were valued as cheaper and more visually brilliant, as Brown himself would recount.[103]

Despite the popularity of synthetic dyes, some individuals and communities maintained their expertise with natural materials. Some Bagobo weavers finished single panels of abacá fabric and submerged them in brews that could yield intense, lasting burgundies. The dyes saturated nearly evenly, leaving some portions almost dark brown in hue where dye likely pooled in a more concentrated manner. These monochromatic panels indicate the intensity of colorfastness and mordant formulations a Bagobo weaver could concoct.[104] For the ikat-dyed fabrics, other preparations yielded striations of color depending on the types of plant material used and the intricacy of the pattern. In Benedict's own account, these differently colored, time-intensive pieces honored Bagobo deities, ceremony rites, marriages, and funerary practice.[105] Benedict herself accumulated names of plants in common use among her informants, male and female. She obtained a medicine box from an elder Bagobo man that contained plants that could help a "woman weave well." Other items included botanical roots for bodily ailments and plant-derived objects to defend against malicious spirits and witchcraft.[106] As in this chapter's opening, plant materials could also assist in preparing woven pieces for unusual transactions by communicating with and facilitating the transfer of a spirit from one fabric to another.

Benedict's acquisitions—both material and informational—did not come without persistence, opportunist dealings, and broken customs. Several of the unfinished skirts in the Benedict collection, for example, indicate disrupted traditions. Cherubim A. Quizon reads these moments as weavers' reluctance to part with garments before they could become the kinds of treasured heirlooms that the ginayán, for instance, had been.[107] After all, what skillful craftsperson would part with a half-finished masterpiece? Weavers shared with Benedict that certain afflictions—such as eyestrain, joint trouble, fatigue, and emaciation—would befall those who neglected to follow the proscription of selling a textile before its completion.[108] Engaging in the sale of an unfinished bag or work of embroidery or lace could "make the women too eager for the society of men." Others hesitated before parting with brass adornments, citing possible sickness or marital discord.[109] In one instance, Benedict persuaded a Bagobo man to sell a brass chain by delivering "'Latin Catholic formulas'" that, she assured, would prevent misfortune.[110] She bargained for her Bagobo informants' objects, including those that had been in their possession for years. Although several declined, others visited her with an intent to sell, and she developed ongoing relationships with particular sellers, approaching them at "opportune" moments, such as after they had lost money at a cockfight.[111]

Amid the emergence of a cash-based economy in the gulf, selling items to an eager researcher was one option for obtaining legal tender. In the early twentieth century, subsistence farming had been a more widespread reality across the archipelago than the mass circulation of cash and coinage, which was more centralized to the colonial capital. US colonists often complained of local shortages of labor to develop urban industries and rural agriculture, but such complaints presupposed a monetary economy that colonists believed native workers would respond to favorably. The uneven development of a cash-based economy meant that labor was hard to recruit, and so, as Greg Bankoff writes, "the solution to wanting more workers was to make workers want more."[112] For the Davao Gulf, as tax impositions and private retail rose, so did the number of wage-earning Bagobo on new plantation land.

The Bagobo became "astute traders," suggests Dacudao, as anthropological pursuits increased competition for Bagobo material objects.[113] Put in this light, the unsewn panels, partially completed pieces, and still-on-the-loom fabrics of

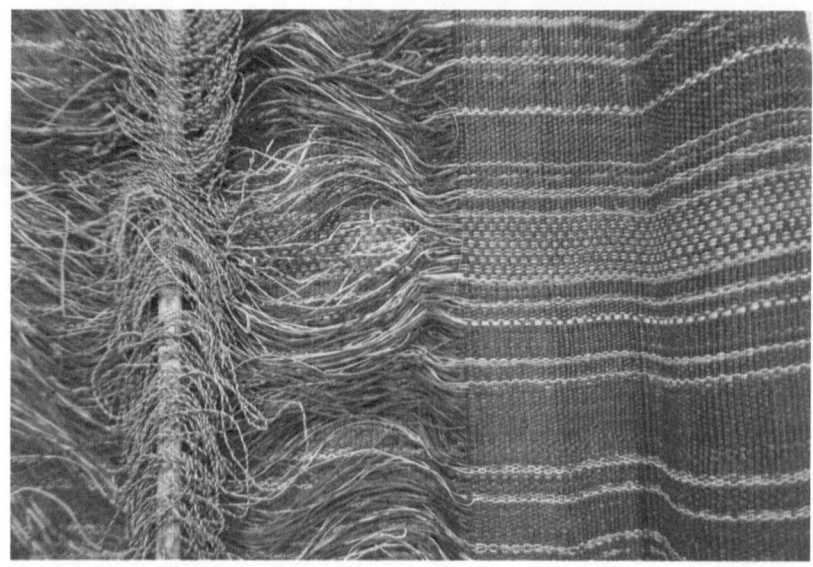

FIGURE 4.9. Detail of "Textile in process of weaving," 70.1/6938 A, Laura Watson Benedict Collection, Anthropology Division, American Museum of Natural History. Photograph by author. Reproduction permission courtesy of the Division of Anthropology, American Museum of Natural History.

Benedict's collection (see figure 4.9) convey the pressures of a changing economic system. Historically, the Bagobo exchanged goods with Moro and Chinese merchants and nearby communities, and finished textiles were very much a part of trade. When a new cash currency displaced traditional economic practices, however, the Bagobo had to adapt the means by which they typically sustained.[114] Benedict was fully aware of this.[115] That Oleng's husband sold her incomplete ginayán to Benedict suggests not only his willingness to entertain Benedict's request but also the forces that made such a transaction vital. From this angle, Oleng and other weavers displayed a flexibility that interrupted, among other things, the artisanal and spiritual equilibrium of their craft.

* * *

Financial duress and mental anguish cut short Benedict's time in Davao. Her quickly evaporating personal funds compounded her difficult relationships with other foreigners in the region. She feared other collectors, especially those with more money, could jeopardize her planned sale to Chicago's Field Museum. Upon her return to the United States, in fact, Benedict struggled to sell her collection. The museum recanted its earlier offer because of the high price she set. Disappointed, Benedict eventually sold her collection to the American Museum of Natural History, where the objects remain.[116] Later assessment of the collection would conclude that of all the types of materials Benedict acquired—from weapons to musical instruments, fishing equipment to utensils—the weaving and textile objects were the most comprehensive of process, material, and kind.[117]

Indeed, Benedict's weaving and textile collection is not only especially rare but also poignantly indicative of colonial transformation. The militaristic rush into the Philippines paralleled botanists,' planters,' and anthropologists' own commercial and intellectual undertakings in Mindanao's south, yet the Bagobo objects evince other ways of knowing that escaped botanists' pages, or even margins. Rather than bemoan colonial botanists' "forgetting" or insist that the science rectify its index fetish—or worse, contemplate the science as responsible for even more encyclopedic preservation—these contrapuntal stories instead put into relief the disciplinary stakes and capitalist moorings of colonial botany in particular and US Empire more broadly in the Philippines. The "primary" work of botany set in motion the "secondary" work of imperial gain.

US colonial botany outfit continued apace in the first decades of the twentieth century. As botanists and collectors scaled Mount Apo, they journeyed to other islands, peaks, and forests to build their repositories. Plant specimens, or their intellectual currency, required energy, resources, and local familiarity. Botanists therefore instantly recognized their dependence on Philippine-born assistants. They also recognized that certain menaces—seen and unseen to them—could hamper their work.

PART III

Assembling a Wider Expanse

5

Field Labor's Menace

> We, too, had our own saints before the Spaniards' arrival.
> I think it would be unfair if God only gave them to the Catholics.
> —ISABELO DE LOS REYES, *La Religión del Katipunan*, 1900

Field laborers commonly joined Philippine botanical expeditions. Foreign botanists hired locally born men for short-term excursions to navigate terrain, carry supplies, and sometimes collect plant material, but these men appeared infrequently in botanists' correspondence and even more rarely in finished publications. Despite a lack of textual documentation of their work, visual records betray this conventional absence.

"Tree fern on Lamao River" (see figure 5.1), from around the 1910s, centers a field laborer, recognizable by his clothing, lack of expedition tools, and bare feet. Exposed roots of the tree to his right point to the sharp, precarious angle

FIGURE 5.1. A once known field assistant posing in "119. Tree fern on Lamao River," Bataan, ca. 1910s, Elmer D. Merrill lantern slides, Archives of the New York Botanical Garden. Reproduction permission courtesy of the Archives of the New York Botanical Garden.

of the hillside with crumbling soil along a tight ledge. A ravine seems to separate the man from the camera operator, and the Lamao River presumably flows below out of sight. The field laborer leans, rather comfortably, against his left hand while crossing his left ankle over his right. Casual, not stiff, he poses in the clearing far from the photographer. Though plant life overwhelms the frame—leaves blur the top right of the image as the camera focuses across the ravine—the composition directs viewers' eyes to the individual under the large fern fronds.

The province of Bataan, where the image was taken, became the site of a forest reserve in 1904, on which the largest US colonial agricultural station was developed.[1] On at least one expedition through the dense forests, botanists scoured the land "on the nature of the country" only to determine that there was "nothing of botanical interest in the region."[2] Paradoxically, their assessments did not slow the frequency of collection in the area. Herbarium specimens from Bataan over a thirty-year period suggest that US personnel closely surveyed the province and its mountain and riverine localities on a near annual basis. A collecting map of the Philippines from 1915 (see figure 5.2) suggests that, in addition

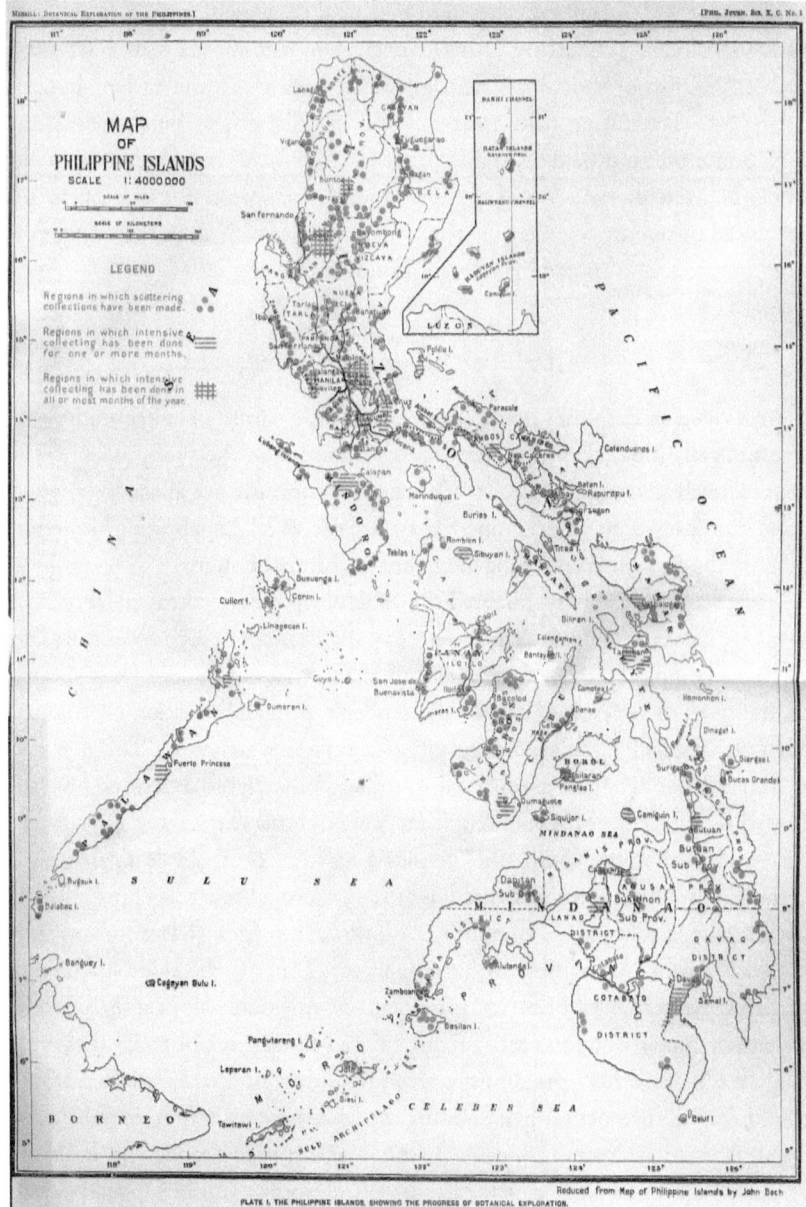

FIGURE 5.2. "Plate I. The Philippine Islands, Showing the Progress of Botanical Exploration," 1915, *Philippine Journal of Science* 10, C. no. 1. Hatched markings on the map denote "regions in which intensive collecting has been done in all or most months of the year." Image and reproduction courtesy of the LuEsther T. Mertz Library, New York Botanical Garden.

to several other locations, southeastern Bataan, through which the Lamao River runs, saw intensive collecting during all or most months of the year.

No mention of short-term field assistance appears in the earliest botanical reports. Recruiting field assistance for expeditions, to be sure, was not easy. Some Filipino field assistants proved intrepid, yet most others were, in Elmer D. Merrill's eyes, "timid or superstitious" and would not investigate uninhabited or remote regions unaccompanied. Identifying an unhesitating Filipino assistant, untroubled by superstitions, was challenging.

The Labors of Colonial Botany

Merrill's assessment raises several questions on the nature of collaboration between locally hired personnel and foreign botanists in the Philippines. What factors motivated collaboration? What sorts of superstitions allegedly plagued most Filipino would-be personnel? How did the word "superstition" function as more than a simple epistemic judgment in colonial botany?

Historians have ably highlighted the vital, complicated role of collaboration in colonial botany.[3] Erasures and intellectual theft surface in such accounts, as do stories of complicity, mutual support, and intellectual innovation. Plant species' centrality to the science of botany can render explicit information on colonial labor tertiary, making collectors and guides seem nearly imperceptible. For a task as accumulative as botanical collecting, certain labor is effaced or remarkably condensed for efficiency of acknowledgment, underscoring the staggeringly uneven nature of collaborative work.[4] The "invisibilizing" quality of science is patent.[5]

Field labor's presence, disguised as it is by colonial recordkeeping, should not be gauged solely by elite scientific visibility. To suggest that labor was simply *made* visible or invisible by a foreign party inherits the value judgments or virtue hierarchies of historical interlocutors. Although "optical metaphors" have their limits, I suggest recognizing instead that field labor was always already visible to other communities, even if readers of scientific publications overlooked it.[6] Momentarily inhabiting the heterogenous vision*s* of field labor illuminates social hierarchy, camaraderie, and intrigue among the ranks of workers. Amid this nuance, perceptions of Philippine plant life appear varied, layered, and conflicting. Similar to what Macarena Gómez-Barris terms "submerged perspectives" that slice through the hegemonic structures of colonial-capitalist extraction, these perceptions reveal a "landscape of multiplicity" that have the potential to "challenge obliteration."[7]

This chapter explores the labor of locally hired personnel of US colonial botany in the Philippines. Professionalization by the end of the nineteenth century heightened the divides among botanical experts. Nevertheless, field guides, plant collectors, draftsmen, illustrators, stenographers, herbarium aides, secretaries, and translators contributed substantially to botany operations. The Philippines, like other once colonized territories, has a storied history of native contributions to naturalist expeditions. German explorers such as Carl Semper (1832–1883) and Fedor Jagor looked to what they termed "diener," or "servants," to complete all manner of tasks, including mollusk collecting, volcano surveying, and ethnological drawings.[8] The Spanish hired insular men to take up salaried ranks as part of the Inspección General de Montes and the Comisión, including rangers, couriers, clerks, and natural history conservators.[9] At the start of the twentieth century, most informal and formal hires were from the Philippines, appointed to assist the Bureau of Science. I focus on field labor, which is distinguishable from other occupations because of its direct ties to expeditions. Botanists also collected in the field but were separate from field guides, porters, and collectors, whose primary expertise in the science manifested outdoors. Often, field labor looms without grand recognition in publications, and only the most prolific of collectors—as measured by the number of specimens collected and kilometers traveled—were named.

The chapter thus begins in the field, where US scientists seeking novelties in the diverse landscapes of the rural Philippines also found themselves afraid of the climate, unfamiliar diseases, uncharted terrain, and sometimes antagonistic local communities. Easier access to precious plant specimens required the services of locally born men. Such recruits enlisted for a variety of reasons: intellectual development, employment, social prestige, supplies, and personal goods could encourage collaboration with foreign personnel.

Photographs from the first decade of US colonization provide a sense of the men whose familiarity with local social and natural environments aided US botanists. Some explicitly trained in the ideals and conventions of botany. Others worked for stints. For the Philippines, the technology of the photograph visualized colonial exploration toward resource exploitation and intellectual investigation. These images reveal *how* colonial cameras were aimed and *what* colonial surveyors valued most. Studying these images can be fraught.

I recommend seeing these images differently to read for the field guides, their know-how, and their "imprint in shaping the visual record."[10] Furthermore, sharpening the contrast between plant life and human actors in these images is necessary to disturb the colonial visual conflation of the floral environment with the field assistants so casually pictured. As Cherubim A. Quizon and Patricia O. Afable note, "heartbreak and irony" tend to follow the unboxing of these sorts of artifacts that imply a wordless display contrived through

US colonial visions of the archipelago.[11] One cannot deploy these images without acknowledging the historical coercion of "forced visibility" or recognizing the "extractive view" that "devaloriz[es] the hidden worlds that form the nexus of human and nonhuman multiplicity."[12] In effect, one might also obtain scientists' and photographers' "subjective investments in the objects of their gaze" and therefore, the "ontology of spectatorship."[13] The successful execution and capture of these stills depended on the field guides' ways of knowing that facilitated the photographic encounter. I focus on their fragmented stories to refuse another "mythic explorer-hero narrative" that envisions colonial botany as a single man's triumph but also to avoid an overstatement of agency difficult to retrace in the scant records available.[14] I analyze these images knowing full well they reveal only part of the picture, so to speak.

As the chapter's opening discussed, not all locally born men worked without pause or were passively dependent on the employment offered by US personnel.[15] "Superstition," as Merrill labeled it, muddied attempts to recruit men to enter forests. English and local Spanish and Tagalog discourse from the time identified superstition, superstición, or pamahiin as a scourge on civilizational progress, Christian principles, and modern education. The word had a derogatory valence usually cast against informally educated, ethnically minoritized, rural peoples of an older, far-off generation, yet several consistently recorded views dealt with particular plant species, environmental formations, and comportment in the forest. Superstition, in other words, was (and still is) an epistemic category.

I examine superstition as it functioned for US colonial botany. In the Philippines, the environmentally oriented rituals, expressions of reverence, and deeply held beliefs it encompassed confronted botany's success in the archipelago. The many sovereign vernaculars of these kinds, their manifestations in practice, and their assortment across the islands were not evenly recorded, universally performed, or assumed by all field guides and collectors. Their textual documentation clashes with the typically oral transmission of such beliefs. Rather than determining the veracity of foreigners' claims against what they cataloged as superstition, I explore their discourse as an impediment to botanical labor power. I argue that for foreign personnel, superstition functioned as an interruption to the physically intensive, extractive labors of botany. For colonial botany, rather than only being an abstract set of beliefs, superstitions were actions that made visible the fragility of field labor and foreign personnel's dependence on locally hired workers. Sovereign vernaculars that fell under the rubric of superstition were engagements with the environment that collided with botany's success. The purported holders of these beliefs and practices unmade the science in the field. At the same time, such engagements helped to delineate the sensible man of science from the laborers presumably outside of reason.

The chapter proceeds through the nature of collaboration between local and colonial personnel and discourse on superstition, weaving in the work of Maximo Ramos (1882–1932), a Filipino plant collector. He began his career as an informally hired laborer before serving as chief botanical explorer for the Bureau of Science. His sudden death in the field in Mindanao after nearly three decades of service catalyzes an analysis of field assistants' own perceptions of threat associated with fieldwork. Claims of superstition could have a flattening effect on how US colonists perceived local labor, yet Ramos's story illustrates how locally hired men could at times act in concert with the colonial botany outfit to safeguard plant collecting and to ensure a steady flow of local personnel.

Perceptions of Locally Mediated Danger in the "Unexplored Field of Labor"

Upon his arrival to the Philippines, Merrill pronounced the islands "an almost unexplored field of labor."[16] Even so, the United States struggled to enlist willing non-Philippine botany recruits to begin to survey the islands' flora. After the founding of the Bureau of Science, Paul C. Freer hoped to populate the Philippines with a "large corps of scientific workers," but in reality "only a few men were actually on the ground."[17] Under the direction of the Second Philippine Commission, Freer allegedly "visited many laboratories" in the United States to identify candidates for positions in the Philippine colonial service.[18] Few trained US men, however, sought to join, leery of the Philippine environment.[19] Even recruiting a competent director for the Jardín Botánico de Manila had proven difficult for the Spanish in the mid-nineteenth century because potential applicants were aware of the little remuneration, the dangers of travel, and the challenges of an unknown place far from home and kin.[20]

Death was common in colonial Philippine botany. The Spanish had cycled through phases of hiring peninsular men to take field posts to replace sick, deceased, or—if they were lucky—retired personnel. Young men with variable training landed in Surigao, Bilibid, Daraga, Bacolod, and Vigan, among other locations. Recruitment posed a challenge as hopefuls responded to openings in *La Gaceta de Madrid* or as higher salaries lured more seasoned candidates, but by the mid-1880s, a reduction of salary and benefits prompted employees to pen a grievance to the colonial government on behalf of the lowest-ranking staff. Their work required navigating roadless terrain, cutting through forests and lagoons, and exposing themselves to diseases in a "burning climate" that made the work "the most grueling" entrusted to any other state functionary.[21]

The sun, humidity, and tropical temperatures reportedly incapacitated US colonists deployed to the archipelago. Complaints of the white body's delicate constitution litter the earliest accounts of US colonization. On a river expedition, US colonial forester George P. Ahern (1859–1942) hired local men to draw and row his boat under the sun for seven hours—a task, he claimed, that "would have killed a white man."[22] Beyond the climate's initial shock, foreign personnel feared tropical illness, having been "moved from their native soil" to a region with unfamiliar contagions.[23] Colonial physicians attempted to create and sport a "white corporeal armature," in a phrase Warwick Anderson coins, that could withstand such new threats, but their vulnerabilities were no less present.[24]

In colonial botany, foreign personnel had been aware of the physical toil they could expect in new lands. Anglo-European botanists' lore on exploration included tales of inclement seas, terminal illness, and inexplicable death. A French fleet in the South Seas under the command of naturalist Philippe-Isidore Picot de Lapeyrouse (1744–1818) vanished in 1818 between Fiji and the Solomon Islands. William Jack (1795–1822), a botanist of the British East India Company after whom the jackfruit was named, died of complications from malaria at the age of twenty-seven in Sumatra.[25] Others, like Celedonio Doñamayor y Moreno (d. 1896), a Philippine forestry assistant, were found dead without clear cause.[26] Such lore warned of locally mediated hostility toward foreign explorers. According to one botanist of Pakistan and the Himalayas, "unfriendliness of the locals" was the "chief danger" in some parts.[27] The euphemistic characterization suggests both the confidence with which some botanists traveled to colonial territories and the precarious interaction with local populations they had learned to anticipate. For a natural science discipline that relied considerably on work completed outside laboratory walls, success depended on facilitating the safe acquisition of plants. Foreign researchers tested how far they would go to achieve a collection, which came with professional spoils, prestige, martyrdom, or an air of sport.[28] In 1810, German botanist Ulrich Jasper Seetzen (1767–1811) learned Arabic, posed as a hajji to collect on the Arabian Peninsula, and reached Mecca only to be poisoned by his local field assistants a year later.[29]

In the Philippines at the end of the nineteenth century, an aghast foreign observer claimed that mothers "in the whole Archipelago" were "teaching their offspring to regard the European as a demoniacal being! an evil spirit! or, at least, as an enemy to be feared." If a white man were to approach someone's home, a "cry of caution, the watchword for defence"—the word Castila for Spaniard, a catchall for whites—would resound, and children would "retreat from the dreaded object."[30] Locals certainly could have perceived foreign researchers with varying degrees of and reasons for suspicion, acting with motivations and intents that troubled easy stereotypes. At the same time, foreign personnel could be aware of their own strange preoccupations in the archipelago: Adolph D. E. Elmer once assured a colleague that Philippine locals would not hurt a "madman," as they perceived him to be, because "no one in his senses would pick little useless plants from tree trunks or from the ground and take them away."[31]

Nevertheless, fears plagued US colonial botanists, especially when suspicions of premeditated violence emerged. In 1908, US forester Harry D. Everett (d. 1908) traveled with two Filipino forest rangers, a Constabulary guard, and a US teacher to the island of Negros. The entire party was allegedly murdered in a plot designed by Ayhao, a man who acted as a local field guide. Newspaper accounts reported that Ayhao and his collaborators drugged the collecting party with the "fumes of [the tuyugtuyug] plant" and killed the full group with allegedly no other motivation than "to kill someone."[32] To botanists, locally hired individuals were the most immediate source of baseline knowledge on plants, particularly useful flora and those with a local name.[33] Other environmental knowledge remained out of foreigners' reach, issuing startling reminders of their own defenselessness in the forest.[34]

Recruiting and Training Local Assistants

Colonial botanists found themselves in shaky alliance with locally hired field labor. Collecting parties brokered agreements with town officials and sometimes directly with male—and infrequently, female—short-term field laborers themselves. These individuals variably stood to gain from working with foreign botanists. Men and women, for example, assisted Sebastián Vidal when he conducted fieldwork in provinces such as Morong.[35] Whether the Spanish hired botanical field labor under the historically compulsory or semi-compulsory labor system remains unclear.[36] In Vidal's case, he traded clothes, beads, and bolo in exchange for safe encampment on excursions.[37] Amid the fledgling monetization of labor in parts of the Philippines at the start of the twentieth century, cash wages meant little in the remote regions where foreign botanists worked.[38]

Nonetheless, some short-term hires could look forward to the rare possibility of more consistent employment, as in the case of Maximo Ramos. His first specimen collections for colonial researchers suggest his fieldwork began in his hometown of Antipolo. Generous specimen samples characterize some of his earliest herbarium sheets such as *Canarium ahernianum* Merr., named after Ahern, who had hired the young Ramos. The then twenty-one-year-old snapped a solid branch of the evergreen tree, mindful of its exemplary flowers, woody outgrowing stems, and leaves still attached by their petioles. To fit the ample material on a single sheet, torn and snipped leaves lay next to cracked and gently folded stems (see figure 5.3).[39]

Ramos's earliest sheets do not refer to him by name. He appears as "Ahern's Collector," a "synonym," as one digital repository obliquely terms it, that masks his labors in service to Ahern's collecting missions in the Philippines.[40] His name only began to appear on sheets—and thus became recognizable to US personnel and the wider botany community—just before the Bureau of Science formally hired him in January 1907.

Photographs of the young Ramos after his hiring feature a poised person (see figure 5.4), bedecked in clothing that usually outfitted foreign colonial personnel and locally born government employees. His job, in many respects, came with social respectability. To be hired by the US colonial science outfit meant a salaried career in government service. Though considered "absolutely untrained" when he was recruited, years into his service he drew the playful envy of foreign collectors made insecure by the quality of his collections.[41] His reputation for obtaining a locality's best and most comprehensive plant samples intimidated others who followed his footsteps in the field.[42]

Ramos would become distinct from locally educated men who trained in a specialized US botany program established in Los Baños in 1909. The College of Agriculture, a campus of the University of the Philippines outside of the colonial capital, prepared students in botany, agriculture, and forestry. Many of the students hailed from landed families across the archipelago.

Since the college was founded at the base of Mount Makiling, students often conducted field studies along its slopes.[43] One instructor related to the university president the importance of field training for young men: "To give the students an idea of how to take care of themselves in the woods, I take each class upon at least three mountain trips each year, one such trip keeping the students out of civilization for three days."[44] Similar to the forestry students who also trained in Los Baños, botany students learned to value fieldwork manufactured through the ideals of US masculinity and rigorous physical labor.[45] Additionally, through an intellectual induction into the science, instructors positioned the young men as within civilization's bounds. This positioning informed assumptions of the types of labor that would then be hired "out of civilization" to assist students on their trips. This social differentiation—along the lines of class, education, and ethnolinguistic group—existed in botany as it

FIGURE 5.3. *Canarium ahernianum* Merr., Natural History Museum, BM000799020, Ahern's collector, no. 422, February 1904, Antipolo. Reproduction permission courtesy of the Natural History Museum.

FIGURE 5.4. A young Maximo Ramos in detail of "No. 171 Narra trees," ca. 1910s. Elmer D. Merrill lantern slides, Archives of the New York Botanical Garden. Reproduction permission courtesy of the Archives of the New York Botanical Garden.

did in the human sciences: not as the expressed categorical objects of study but as reflected in its professional ranks.

Images of the earliest College of Agriculture students include shots of young men in forested and provincial settings. These men are typically dressed in shoes, collared shirts, and ties, distinguishing them from field assistants hired

FIGURE 5.5. The original caption (not included here) of this photo reads, "Cargadores April 9 1914 Sariaya," Philippine Photos 2201–800, box 1, Frank C. Gates Papers, Bentley Historical Library, University of Michigan. *From right*, a field assistant with College of Agriculture students N. Catalan, E. Quisumbing, and V. Sulit. Reproduction permission courtesy of the Bentley Historical Library, University of Michigan.

in situ, who in Spanish and US colonial-era illustrations and photographs are barefoot with collarless shirts.[46] In the image captioned "Cargadores" (see figure 5.5), three College of Agriculture students stand to the right of a field assistant, likely hired in Tayabas province. Field assistants often served as cargadores or stevedores during field expeditions. Hours of trekking necessitated careful carriage of essential equipment like wooden presses, blotters, mounting material, reference publications, rations, and camping supplies. At the start of the twentieth century, Filipino cargadores, who worked the docks of Manila and Cavite, earned 40 to 50 cents a day with a one-hour break.[47] How much botanists paid expedition cargadores in the Philippines is uncertain. Frank Caleb Gates (1887–1955), a botany professor at the College of Agriculture, may have assigned the students in figure 5.5 to function as cargadores for the trip to Sariaya while using the plural "cargadores" ironically.

Formal botany training, publications, and herbarium curatorial skills circumscribed professional expertise in the Philippines, an extension of wider currents of professionalization across the science in the United States. Ramos ostensibly could not, over a decade into his service, identify a full family of plants.[48] When

Merrill named a genus of the Poaceae family *Ramosia* in 1916 to honor his collecting work in Sorsogon, Merrill mentioned that some of his material was still "inadequate," lacking mature flowers and fruits to help determine species. Publishing the paper with the material collected by Ramos, who had "no botanical training," meant that the available specimens remained substandard.[49]

The Menace of Superstition

Ramos's lack of formal botany schooling did not preclude him from relative repute in his career. A reduction in bureau staff in 1920 demonstrated that his work was indispensable to the institution, such that Merrill subsequently recommended him for a promotion with an almost 70 percent raise in salary. Upon promotion to chief botanical explorer, Ramos was described as "exceptional" compared to his contemporaries. Even without training in systematics, he could visually discern common from rarer species and would dedicate time to spotting less familiar plants to build the colonial herbarium. He was, at the time, the only dependable employee who could obtain "material in bulk" with economic potential as well. On Ramos's aptitude, Merrill extolled, "He is absolutely fearless in visiting remote and sparsely inhabited regions. He is not troubled by superstitions regarding mountains and forests, and on each field trip that he makes he always proceeds to the most interesting botanical region in the area to which he is assigned regardless of the distance or the difficulties involved in reaching his destination, and is inured to the hardships incidental to field work in remote and sparsely inhabited regions."[50]

To Merrill, Ramos could spot plants, collect relentlessly, and tread without superstitious fear—the qualities of an excellent "field man in botany." Merrill's characterization of other field labor as "too superstitious" was part of a larger ecosystem of critiques wielded against superstition, which US colonists reported as a common nuisance. Freer disparaged the general Philippine populace for holding superstitions that fostered "disbelief in measures which the modern scientific world has recognized as necessary."[51] This appraisal was neither novel to the twentieth century nor unique to foreigners. Detractors commonly placed the word in contradistinction to the ideals of Christian faith and Anglo-European modernity.

Catholic missionaries cataloged the archipelago's numerous belief systems upon their arrival. Their records translated such beliefs into frameworks, according to Vicente L. Rafael, which an Iberian Christian audience could understand. The Tagalog *nono*, for example, puzzled missionaries, who translated the term as a tutelary spirit as understood through the theological and historical valences of paganism. In the local sense, *nono* "always referred to something more than could be spoken of: spirits in nature and ghosts of dead ancestors," with no fixed names or palatial mythologies.[52] The very act of writing down the belief systems also came with a decontextualized appeal, divorcing the belief structure from the moment of its utterance for future commentary by a religious acolyte, an ethnologist, or an observer. Through the nineteenth century, religious and secular Spanish commentators such as Wenceslao E. Retana (1862–1924) inventoried the most common superstitions among Tagalog and Visayan communities, many of which appeared in ilustrados' writing as well.[53]

Trinidad Pardo de Tavera had railed against superstitions that posed medical harm to pregnancy and childbirth. He noted how many feared the patianac, a vampiric creature that took the form of an infant under trees to attract unsuspecting adults.[54] Writing in 1915, Fernando R. Calderon, an obstetrician and first director of the Philippine General Hospital (founded 1907), mourned provincial obstetrics that had "no influences, such as exist[ed] in Manila, to abolish superstitious ideas concerning midwifery."[55] Some elites considered "superstición," "pamahiin," or "pagtuotuo" to be the "worship of false gods," relegated to a far-off time.[56] For others, they provided a lens to view a Philippine life that was distinct from European persuasions.

Isabelo de los Reyes featured superstitions in *El Folk-lore filipino*, abbreviating several Ilocano customary traditions. For example, "It's ill-advised to throw stones or other objects or to point at trees in remote locations," he wrote, "because maleficent spirits may punish us by keeping the throwing or pointing arm stiff and unable to bend."[57] He elsewhere deployed the term "anitismo," from the word anito—distinct from animismo or animism and the successive emergence of Hindu, Muslim, and Christian beliefs in the Philippines. To de los Reyes, anitismo was the original Philippine religion found among the Itneg of northern Luzon and within the practices of the Tagalog, Bisaya, Bikol, Pangasinense peoples, and other "civilized Filipinos." It was the worship of anitos or spirits of ancestors, seen most generally in the veneration of animals, mountains, rivers, astronomical phenomena, and other beings of nature.[58] As Megan C. Thomas explains, de los Reyes wrote affectionately of the "incredulous, superstitious peasant and the particularities of local culture" to the extent that he also believed they could acquire a kind of civilizational modernity already enjoyed by more Hispanized or cosmopolitan Tagalogs, Ilocanos, and Visayans.[59] Indeed, such affection appears in the epigraph to this chapter, as de los Reyes quips over the collection of saintly persons that God has erroneously given only to Catholics. Pedro Paterno's *Nínay* spent a considerable number of footnotes detailing the aswang of the Visayas, the dwende of forests, and the powers of the anting-anting, talismans carried for their protective influence, at times interchanging "tradición" (tradition) with "superstición."[60]

Social divisions emerge in local catalogs of the turn of the century that throw light on who allegedly practiced superstitions and where they were held. These divisions could be emblematized in the text-artifacts themselves: to write of superstition recalled an earlier mode of distanced observation initiated upon European contact, which could unevenly distribute epistemic authority for foreign audiences.[61] Such writings further reflected educational attainment, moral purpose, Christian baptism, and ethnic difference. The trilingual periodical *Patnubay nğ Bayan* (The People's Guide), featured columns that referenced pamahiin in Philippine society as antithetical to Masonic beliefs, for example, and common to peoples outside of urban centers.[62] *Pagsulong sa Karunungan* (Advancement through Wisdom) prefaced a 1911 issue with the headline, "The New People," who were to be the "enemy of idiotic superstitions or old beliefs."[63]

Some Tagalog writing, however, reminded its readers that superstitions were not exclusive to the Philippines nor altogether abominable. One thought piece on farming pointed to farmers elsewhere in the world who were historically wont to pay homage to gods and beings that could help with harvests.[64] Popular astrology was folded into superstition too. The regular entertainment column, "Hula o pamahiin ng mga Astrologo" (Astrologers' forecast or superstition), associated dominant personality traits or portents with each day of any given month. If one were born on March 14, they would be lovable; on April 27, rather dull.[65] For some Tagalog writers, these beliefs warranted a grain of salt and reverent reasoning for the superstitious "older generation," rather than be cast as absurdity. To teach the purpose of a semicolon, one Tagalog grammar offered the following example: "Let us not fully listen to and believe our elders' inventive stories and superstitions; but if there are good lessons and important teachings let us listen intently with our hearts."[66]

During the US colonial period, superstition had no single consistent definition, though it often referred to beliefs in the power of relics, nature, humans, and spirits. To some US colonial functionaries, superstitions were abstract beliefs that contradicted "a morality-cultivating faith in the Christian god" and therefore provided reason for a "religio-racial governance." In their estimation, superstition had ties to banditry and messianic revolution, which required militarized containment.[67] Certain beliefs emerge consistently throughout US records, many dealing with the natural environment forming a class of "environmental spirits" as one modern Philippine ethnographer has termed them.[68] These recorded beliefs conflicted with a non-vitalist perspective of plant life and of nature writ large.[69] As such, the very construction of nature for some involved the presence of spirits—an onto-epistemology discordant with superstition's deepest critics.

Species could house maleficent, generous, or playful beings. Formations could be home to deities. With respect to nono, locals reportedly sought consent when obtaining flowers or fruits from trees, asked for passage along creeks and groves, begged pardon when cutting timber, and excused themselves "by saying the priest ordered it."[70] The buso of the Bagobo, as Laura Watson Benedict wrote, widely haunted the environment: "they people the air and the mountains and the forests by myriads; their number is legion."[71] Balete trees reportedly elicited a certain amount of trepidation in passersby. In one catalog, "Tagalog companions" on a foreigner's hunting expedition refused to sleep beside the tree's engulfing aerial roots. When passing the tree, the companions would address it as though "requesting permission of a superior to pass."[72] The same behavior occurred in Sulu.[73] Dean Worcester recorded what was likely the balete, which his team had been told was inhabited by "white men" and whose branches would bleed if cut.[74] The plant was so frequently referenced on field expeditions in Northern Luzon that Merrill eventually named a local species *Ficus balete*.[75]

The photograph "Balete tree on Tuai" (see figure 5.6), which also appears in this book's introduction, shows a field assistant lounging on a balete as it ensnarls a tuai. Without knowing what transpired when he approached the balete and climbed its roots, viewers can readily observe his embodied knowing of the tree. He rests his back against the aerial roots, aware of the balete's sturdiness. The image centrally features the balete's parasitic hold of the tuai, a tree critical to US colonial timber prospecting and once deemed "the worst enemy of the forest" by surveyors because of its ability to strangulate other profit-bearing trees.[76] This image depicts then not only a plant understood to host spirits—and a field assistant's physical comfort around it—but also the problem of superstition in its clearest forms: as a hindrance to resource extraction, as an obstacle to active field labor, and as a reminder of foreign researchers' dependence on local personnel toward botanical progress.

Through one interpretive frame, field assistants' belief systems could interrupt the unfettered surveying and collection of the Philippine environment. In 1911, a colonial surveyor wrote of the local alcohol industry, narrating his field research in Tayabas to sample buri palm sap used in the manufacture of spirits. He related how natives believed the buri would burst "if a woman touches or interferes with it in any way while the sap is flowing." The surveyor offered to pay the wife of one of his male field assistants ten pesos—understood to be a "large sum" to her—to climb a ladder to reach the top of the palm in order to "see the phenomenon." She refused. Her husband also argued against the request.[77] One cannot suggest all recorded claims of adherence to superstition were intentional acts

FIGURE 5.6. "Balete tree on Tuai," no. 136, ca. 1910s, Elmer D. Merrill lantern slides, Archives of the New York Botanical Garden. Reproduction permission courtesy of the Archives of the New York Botanical Garden.

against unregulated botanical collection or surveying. Yet superstition interfered with colonial botanical work on the whole and, quite literally, its material gains.

Superstitions could further operate as a form of reverence, tied to a vision of the landscape inaccessible to foreign and elite critics. As Erik Mueggler describes of the Nvlvk'ö collectors who assisted Scottish botanist George Forrest

(1873–1932) in China, they worked "through strata of the landscape invisible" to foreign botanists. For the Nvlvk'ö collectors, these were ancient paths charted in local history, distinct from "a new botanical geography" of Yunan established by Forrest. Consequently, "two archival regimes" emerged during the expedition: one tied to Forrest's botanical investigations of the region, and the other to local historical narratives of the ancient landscape.[78] Would-be field assistants who refused to enter forests or enter unaccompanied could render a different landscape: one that recognized the anito of Mount Makiling, the buso of Mount Apo, or the benevolent deity of Mount Arayat.

Through another frame of interpretation, superstition may be one of many tools that field assistants used to avoid tasks they did not want to complete. Anecdotal accounts of field guides and informants' displeasure with foreign plant researchers are common in the history of botany. Made-up plant names and inaccurate directions circulate the tomes as do instances of direct labor refusal.[79] Throughout her collecting career in Southeast Asia, Mary Strong Clemens (1873–1968), a US plant collector, remarked on the local people who aided her through her collecting trips. While in the northern Philippines, she complained that a field assistant had "gone on strike" as her collecting group approached the base of a mountain.[80] Foreign collectors could promise goods and food, but assistants could stiffen at what was expected of them. Clear communication of expectations was never guaranteed before or during a trip. That an assistant went "on strike" was not surprising given the conditions of collecting. What could be expected of an assistant could outweigh the supposed reciprocal benefit on collecting missions in the Philippines and elsewhere.[81] Clemens narrated how a rod in her husband's hand "straightened out matters" before others could join the recalcitrant assistant, and such a physical instantiation of authority was one of several the couple acknowledged in their years of collecting.[82]

Not simply an abstract set of beliefs considered an intellectual irritation, superstitions upheld a particular sovereignty in the face of scientific extraction. Workers could choose to deploy what were considered superstitions at their will, mucking up the exploits of the foreign personnel. The sovereign vernacular, therefore, emerges by way of workers' decisions to practice their beliefs or to use them as a reason for labor refusal—or both. As Londa Schiebinger writes of early modern European naturalists, "the developing conventions of colonial master/servant relationships rendered Europeans unduly dependent on their guides in the tropics."[83] Into the early twentieth century, this dynamic persisted: not only in the lack of recruits for fieldwork but in the increased sense of a way of knowing plants that made it hard to achieve headway. Local collaboration was essential to survival because of menaces—depending on whose vantage point—on fieldwork.

Maximo Ramos: Contrapuntal Discourse on Threat

I turn back to Ramos (see figure 5.7) to explore how locally hired field labor may have viewed the menacing danger of botanical fieldwork. Some Filipino personnel were well aware of its threats, acting so as not to disrupt scientific operations. Merrill's recommendation in 1920 indicates that Ramos continued his career with the Bureau of Science. During this time, Ramos wrote to Eugenio Fenix (1883–ca. 1939), a fellow plant collector, mentioning at times that he would someday want to pursue agriculture because the hardships of botanical work were unnecessary to endure without an heir, presumably to enjoy the spoils of the income he received.[84] Through the 1920s, Ramos participated in botanical missions to Bohol, Isabela, Cagayan, Mindoro, Balabac Island, and the Batanes group.[85] By then, he was also known as part of a collecting pair. He completed many of his trips with Gregorio Edaño (1896–1960; see figure 5.8), another plant collector who began his work with the bureau in 1916.

During a collecting trip in 1932 with Edaño and other assistants to Mindanao, Ramos fell ill while in Buayan, Cotabato. According to a memo from the Bureau of Science to Merrill, Ramos died of heart complications from malaria after a brief illness during the collecting trip. Due to limited facilities and personnel, Ramos's collecting partners left his remains in Buayan. Sometime after notice of his death reached Manila, bureau employees were directed to disinter his remains and bring them back to his widow.[86] Soon following news of his passing, Merrill was implored to publish a brief obituary in *Science* to commemorate the tremendous loss to Philippine botany, heralding Ramos as one of the "few individuals of any country [to] have prepared such an extensive series of herbarium species."[87]

After Ramos's death, Edaño continued his career in collecting and in teaching. He was responsible for training new groups of botanical collectors associated with the Philippine National Herbarium in the mid-twentieth century. During a field research conversation with me, one of Edaño's former students commented that Ramos's colleagues kept the details of his death secret from most people within and outside of the bureau. Edaño recalled Ramos having developed severe gastrointestinal pains. One of the collecting group's options, Edaño confided, had been to create a kind of enema to help Ramos evacuate whatever bacteria he had in his gastrointestinal tract. Without access to medi-

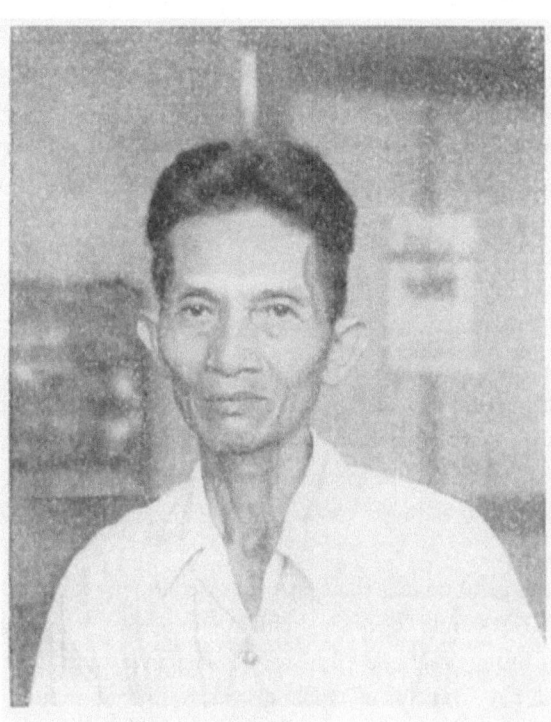

FIGURES 5.7 AND 5.8. Maximo Ramos (*left*) and Gregorio Edaño (*below*) became a regular expedition pair, collecting between them thousands of specimen numbers from the Philippines. Images from *Flora Malesiana*, vol. 1, ser. 1, edited by C.G. G. J. van Steenis et al. (Jakarta: Noordhoff-Kolff, 1950), courtesy of the LuEsther T. Mertz Library, New York Botanical Garden.

cine or medical facilities, however, the group failed to treat him successfully. Out of necessity, they left Ramos's body, which was exhumed at a later date. Edaño and others feared that the bureau would receive backlash from the public should word spread regarding the dangers associated with botanical collecting. In an effort to reduce outcry, especially from Ramos's widow, malaria was cited as the cause of death. This version of events, the former student explained, would not deter local men from botanical collecting, unlike rumors surrounding illness, death, and disease of unclear origin.[88]

Several screens of distortion exist in the shadow of Ramos's death. These distortions affect the nature of death within the institution proper, for the general public, in the confidence between a teacher and his pupil, and for the historian's probe of the archive. In the first sense, the circumstances of Ramos's death perhaps could not be communicated even to Merrill, who at the time had already been employed outside of the Philippines at the New York Botanical Garden. Between Edaño and the other assistants on the expedition to Cotabato, tracing who could attest to Ramos's final hours is difficult. If indeed Ramos died as result of an unspecified gastrointestinal illness and not malaria, one must infer that for the general public at the time, to perish because of malaria, a common scourge, was more acceptable in botanical collecting than more indeterminable ways of dying. Regarding Edaño's account of Ramos's death, the anecdotal quality of the experience might have functioned not only as a pedagogical tool but also as a divulging of secrets. These distortions sustained the institution's ability to engage labor power and prevented what might have been a public disaster had Ramos's widow known the full circumstances of his death. Certainly, something must be said of the widow's perceived influence. Without her assumed capacity to damage the institution's reputation, such distortions may not have been necessary.

Some Filipino men assisted the scientific outfit, careful to reduce public perception that may have impacted the recruitment of future personnel. These details were seemingly kept secret not only from would-be hires but possibly from Merrill and from other official records. Locally hired workers had been well aware of the dangers of botanical exploration and the hardship of recruiting willing men to engage in potentially perilous activities. Their possible intent offers a different side of the discourse on the labor power of colonial botany, one that provides a contrapuntal reading on the fragility of field labor.

* * *

I acknowledge that my choice to ponder the deaths of several men reenacts a privilege that expects that lives should "be made useful or instructive, by finding in [them] a lesson for our future or a hope for history."[89] However, to exhume their stories and to "create memory" are acts of putting "the body of history back together" not least of which for a diverse class of laborers and for those whose untimely or violent deaths continue to rattle practitioners of botany today.[90] To write of them is to confess the weight of examining a particular dimension of history of science. Ramos's case, in particular, challenges simplified accounts of labor problems and complaints, many positioned against Filipino workers for disruptive acts. Claims of superstition could simplify how US colonists perceived local labor, whereas Ramos's case study demonstrates how locally hired men could, at times act in concert with colonial botany to safeguard plant collecting and maintain an inflow of local personnel.

I close with a return to the word superstition, a term with Latin and French derivation. Its etymological root, *superstitio*, began as a neutral term, "reserved for strange, nontraditional activities."[91] Yet, by the end of the first century, it became a pejorative to identify individuals who had sacrificed their dignity in service to deities through exaggerated rituals and adherence to prophecies. Likely influenced by the Greek philosophical treatment of *deisidaimonia* (literally "fear of demons," or excess religiosity), the Romans used *superstitio* to refer to exaggerated dread of the gods and to foreign religions, such as Judaism or Gallic animism.[92] Like his contemporaries, Pliny the Younger (61–ca. 113 BCE) considered particular Christian practices a threat to Roman society or a "contagious disease" to isolate.[93] Superstition, in today's sense, emerged in Middle English in the fifteenth century, during a period of Norman linguistic influence. In a way, superstition, a word rooted in Latin, was a longtime shared snub to vernacular expressions of belief, one that—by the late nineteenth and early twentieth centuries—Spanish, Tagalog, and English speakers alike could recognize and commiserate over. In other words, it was a common slur, rooted in botany's chosen universal language, against a way of knowing plants.

This lexical history differs from pamahiin or pagtuotuo, whose associations with superstition, I surmise, appeared during the colonial period. On their own, the root words "hiin" and "tuotuo" do not immediately imply supersticíon or superstition, words that would have been introduced during Spanish and US contact. Pamahiin derives from a combination of two prefixes associated with the root word "hiin": "pang" as in "intended for," grammatically transformed into "pam" and "pahiin" as in "leaning toward hiin." Hiin, according to some Tagalog speakers, can relate to the act or notion of removal. Tuo can be the root word for believing or heeding, but reduplication—as in *tuo*tuo—leads to the diminutive form of the word, as in smaller or lesser beliefs. Their modern Spanish and English direct translations as supersticíon and superstition, in other words, may be viewed as colonial introductions.

Colonial botanical fieldwork in the Philippines presented threats that foreign researchers could not immediately apprehend. Collaboration with local personnel was key, but superstition impeded seamless, continual work. Superstitions could shroud any number of reasons not to enter a forest or to snap, slash, or pluck plant material, yet they also functioned as a foreboding reminder of colonial botanists' vulnerability during collecting work. Such vulnerability was not lost on local personnel.

Collaboration continued to be key in the developments of Philippine botany, and eventually, wider regional phytogeographic studies in Merrill's own work and that of the US collector Clemens. The following chapter details how Merrill, a devotee of the professionalization of botany and critic of superstition, experienced his own change of heart regarding the linguistic stability of new imperial botany—a transformation he underwent while in the archipelago.

6

The Latin Babble

In many cases such names are more constant than our supposedly "permanent" Latin binomials.—ELMER D. MERRILL, Manuscript of commentary on *Flora cochichinensis*, 1919

Mary Strong Clemens's botanical collecting exhibited unremitting attention. A graduate in the natural sciences and a self-trained collector, Clemens had an unusual energy for detailed work that complemented that of other collectors employed at the Bureau of Science. She tried to "paint the best mental picture possible of the flora" in any given locality. However, collecting alongside the likes of Maximo Ramos, with whom she worked closely, spurred insecurity. To her, following behind him on any botanical expedition was lofty, knowing he had already meticulously scoured an area. In her words, she assumed the best collectors would leave handsome plant material "for any wayfaring old lady to add at least a species or two" to an herbarium set.[1] Not only did Clemens's livelihood depend on it, so did the careers of botanists like Elmer D. Merrill.

In 1926, when she was hired to complete a botanical trip to central Indochina, Clemens collected, fastened, and labeled 1,200 specimens, which she sent to her botanist-collaborator. Merrill had been working to revise a flora of Indochina and relied on her specimens for the revision. Thoroughness, especially with local plant names, was the order of the day. After her trip, the collections arrived in California for Merrill in spotless condition, but her labels, Merrill implied, were "'so worthless the writing would soon be effaced.'"[2] Distressed by his appraisal, Clemens defended herself, "With *scrupulous care* I put labels on *every species* and would consider it no less than dishonest had I done otherwise."[3] The seemingly minuscule part of the entire collecting set was crucial: such labels carried local names collected by Clemens—names that would inform considerations for international taxonomic practice in botany.

A Vernacular amid the Babble

Typical collecting labels include the location of a collection, a preliminary identification of a species, and a plant's morphological features. Collecting local names was not required of colonial botanical fieldwork and even today is not a requisite practice across all institutional programs in botany. In the first third of the twentieth century, plant knowledge of a community, of guides, and of those present on expeditions did not always find its way into collecting notes or as part of the finalized voucher for an herbarium accession. In contemporary botanical practice, herbarium researchers rely on vouchers to tease through the seeming limitations of local names to reach the certainty of taxon identity.[4] In this chapter, I examine a moment in the history of colonial botany when the reverse was the case: when collecting labels and vouchers *with* local names enabled taxonomic certainty. I thus come full circle, back to the beginning of this book. Instead of the predicament of varieties, however, I show how a sovereign vernacular—that of common nomenclature—made botany.

A project born of Clemens and Merrill's collaborative career hinged on the nomenclatural vernacular: a 1935 commentary on Portuguese Jesuit botanist João de Loureiro's (1717–1791) *Flora cochinchinensis* originally published in 1790. By the 1930s, Merrill had written reams on Indomalaya, a wide floristic realm extending from what is now considered the South Asian subcontinent, southern China, mainland and island Southeast Asia, and up to the Ryukyu Islands. Botanists today configure the realm's expanse based on specific family spread, such as that of the Dipterocarpaceae. To achieve such far-reaching work, Merrill revised Georg Eberhard Rumphius's *Herbarium amboinense*, Manuel Blanco's *Flora de Filipinas*, Nicolaas Laurens Burman's (1734–1793) *Flora indica* (1768), and the Malesia works of Dutch naturalist Maarten Houttuyn (1720–1798). Merrill's publication on Indomalayan flora advanced US intellectual claims internationally—claims made more substantial because of specimens obtained by Filipino and other foreign botanical collectors without whom he could not have completed his publications.[5] His commentary on *Flora cochinchinensis* marked an important moment in regional botany studies that had been gaining momentum since the nineteenth century. All the while, he rose to become one of the most prominent leaders of the International Botanical Congress (IBC).

Clemens and Merrill's collaboration shows the endurance of local plant names used not only to correct so-called pioneer botany works of the past but also to challenge Latin binomials as the allegedly stable nomenclatural system of the science. To Merrill, long-used non-Latin names threw light on the precarity of Latin nomenclature, its inconsistencies across time, and botanists' disordered frenzy to designate different Latin names to the same species. Therefore, investigating the history of botanical nomenclature not only reveals how Anglo-Europeans evaluated others' knowledge systems.[6] It also reveals how botanists evaluated their own.

Far from being a perceived asset, multilingual scientific communication tormented the earliest European naturalists seeking a single shared tongue.[7] For botany in particular, the "Babel" of plant names, as it has been characterized, racked pre- and post-Linnaean researchers.[8] In China and in South Asia, flora, herbals, and materia medica compendia listed local synonyms of plant names, like Li Shizhen's *Compendium of Materia Medica* published in 1596, which includes a "plant name map" that tied phytonyms to their locatable use and provenance. Before the emergence of Linnaean nomenclatural norms, botanical tracts in Europe displayed a variety of languages, such as Hendrik van Rheede's *Hortus indicus malabaricus*, which contains Malayalam, Arabic, and Konkani. Even with modern botany's institution of Latin binomials after Carl Linnaeus's *Species plantarum*, the way botanists deployed Latin was not wholly consistent.

As much as Babel could be projected as the figurative chaos that confronted those attempting to order the world, so too was Latin imagined as a unifier. Before Latin's ascendance as the Western European language of science, it had been subordinate to Arabic and Greek. Arabic had been the foremost language of natural philosophy, and the medieval period witnessed a translation enterprise of Greek tomes into Arabic. In the twelfth century, translation of Greek and Arabic tracts into Latin—spurred by the arrival of paper-making technology from China by way of the Arab world—gave a then-small community of scholars the means to produce scholarship in Latin. An ecclesiastical language and the idiom of the state, Latin obtained exalted status during the Renaissance when local languages emerged across Western Europe for daily administration. Despite its potential as a universal tongue, few could actually access Latin for scholarly means. Its users confronted the dominant scientific languages of Sanskrit and Classical Chinese with their rich textual traditions. Latin's reign as a lingua franca for knowledge was more fleeting and more challenged than has been remembered, considering that intellectuals had their own "Scientific Babel" with which to contend up through the nineteenth century, when a handful of languages became commonly used for scientific production.[9]

This chapter spotlights a moment when the non-Latin exposed what I call the Latin babble. I twist "Babel," an oft-used pejorative characterization of the multitude of plant nomenclatures that remain(ed) out of reach to the global few. Defined in a nonbiblical sense as "a continuous low or confused sound, especially the sound of several people talking," babble captures how botanists struggled to find consistency and standardization through Latin binomials.[10] Botanists believed themselves to be using the same universal language, but as this chapter shows, the racket of identifying and naming species contributed to thick noise, making the work of correcting early botany tracks impossible without the local plant names. Imperial botany's race to dole out Latin names contributed, among other things, to synonymy, which occurs when binomials have been reduced as synonyms to what becomes considered an accepted scientific name. Synonyms pile up when botanists name—and subsequently rename—the same species, sometimes unknowingly, or sometimes in the miscomprehension of an earlier botanist's work. Because botanists identify, name, and revise identifications across space and time, synonymy captures Latin babble's historical and geographic dimensions, or as Alphonse Pyramus de Candolle once wrote, "the history of the science."[11] By the first third of the twentieth century, international botanists envisioned finally rectifying the problem.

This chapter begins with a review of Merrill and Clemens and what their Philippine-born interlocutors taught them about the environment. For Merrill, who trained in the temperate United States, the Philippines presented a challenging landscape. For Clemens, identification work in Latin and in local nomenclature were equally important to her and to her reliability as a collector for foreign patrons. Their Philippine training primed the choices they made toward the publication of Merrill's 1935 "A Commentary on Loureiro's Flora Cochinchinensis." I detail the process by which Merrill revised Loureiro's flora while relying on Clemens's field research. His revisions coincided with his own leadership on the IBC. The type of nomenclatural diplomacy he modeled reflects critical changes in his manuscript and in international botany at the time.

Merrill's work suggests how plant names were cataloged and used, often among Anglo-European botanists conversant principally with each other, and how colonial social dynamics during fieldwork conditioned their entextualization, or the process of rendering discourse as text.[12] Even if names could endure, colonial social structures influenced the entextualizing of what could be deemed fixed local names. Recall the sampaguita, the subject of Chapter 3, and its other recorded monikers in the Philippines, such as hubar among Tausūg informants, kulatay among Kapampangan speakers, lumabi in Maguindanao regions, or sampaga wakat-dangka for some Mangyan.[13] In published works, these identifiers have been transmitted through a kind of authoritative transcription at botanists'—or even the historian's—discretion.

The Philippines and Merrill

Merrill's ambivalent relationship with Latin began at an early age. Born to a modest agricultural family in Maine, he did not plan to follow the academic path of which the classics were an integral part. For much of his youth, his family's limited means foreclosed the prospect of advanced education. He instead busied himself with regional fowl, rocks and mineral deposits, and woody perennials, identifying plant specimens by looking up common names. Along with his twin brother, Dana, he did not consider serious study of Latin, having instead elected to study English—a more practical language, in their opinion—in high school. Decades into his botany career, he regretted the decision: Latin was not only essential for systematics but also for entry into Maine's top private colleges. Without preparation in the classical language, Merrill had to study at a state school and remained, as he phrased it, self-trained in botany, "a hard school, and one that supplied a most sketchy, inadequate training" for the career that followed.[14]

In college, Merrill took survey biology courses, hiked the New England countryside looking for plants, and developed a private herbarium. While studying, he took the US Civil Service exam to become an agrostologist for the Department of Agriculture. The exam included translation of some botanical descriptions and, fortunately for Merrill, no Latin grammar. His success brought him to a research station in New York, and then to Washington, DC, in 1899, where he worked in systematic botany.[15]

The Philippine–American War transformed Merrill's trajectory. His twin brother joined the US military as it confronted local resistance in the Philippines. Merrill was among the dozens of US scientific workers sent across the Pacific.[16]

He twice declined recruitment, professing his ignorance of Philippine—let alone tropical—botany, but his supervisor at the Department of Agriculture assured him, "Nobody in the United States knew anything about the Philippine flora and that [he] had just as good a chance as any one."[17] The twenty-six-year-old Merrill came to the Philippines in 1902 and spent the next two decades there. Ready to apply the knowledge he gained in the United States, he arrived at the "infinitely more complex and difficult task" that was Philippine botany.[18] The environment hosted natural life unlike that seen in US climes. Merrill offered a bleak assessment of the botanical work completed on the archipelago prior to the US arrival. Beyond crediting Sebastián Vidal for his publications and Regino García for being "one of the very few natives" with a record of botanical work, Merrill wrote, aghast, "It is doubtful if any country in the world of a similar size has such a high per cent of 'unknown' described species as has the Philippine islands."[19]

Merrill was quick to suggest that the Philippines lacked locals who could competently pursue taxonomic work, yet he soon realized that understanding the Philippine environment required collaboration with native personnel, trained or untrained in colonial botany. He relied on native assistants and their familiarity with Philippine languages. His many critiques lobbed at the Spanish notwithstanding, Merrill referenced the botanical writings of Ramon Martínez Vigil (1840–1904), Blanco, Vidal, and García, which offered indices of local names. Presented with an archipelagic landscape of, at the time, an estimated eighty languages, Merrill recognized the "unending" task of cataloging names.[20]

During this time, colonial botany had been slowly pivoting since the nineteenth century toward more region-wide investigations of plant material. Merrill built a considerable part of his career on flora revisions within this new orientation, as opposed to the colonial borders that confined many of his predecessors. Likewise, in Japan, botanist Takenoshin Nakai (1882–1952) had proclaimed an "East Asian systematics" founded on studies linking colonial Korea, China, and Japan. Nakai sought to establish a new imperial center that could go toe-to-toe with Anglo-European botany.[21] For the Philippines, US colonial botanist Harley Harris Bartlett (1886–1960) celebrated this "reaching out to adjoining regions" for which, he claimed, there had been "no political motives, but merely a carrying out of the natural impulse to complete our scientific knowledge by correlating the botany of the Philippines with that of her neighbors."[22] The statement, however, elides the international posturing of US botanists who sought to take the mantle from imperial botany centers in Europe.[23] These new regional correlations were still conducted on colonized terrain, but an international attitude suffused the science. For these botanists, regional botany seemed to be (optimistically) prescribed by nature and not by imperial politics.

Merrill set out to correct older botany works and began to regionalize flora of Indomalaya broadly, which required reading relevant successor publications. He observed how publishing naturalists changed or accepted Latin binomials as they pleased.[24] They irregularly deployed descriptive Latin, made typographical errors, and did not maintain herbaria from which they would have developed their descriptions. Amid the cacophony, local names offered Merrill the first resonant clue to identifying species. For instance, in his botanical notes on *Flora de Filipinas*, Merrill wrote of naturalist-friar Antonio Llanos (1806–1881), a close collaborator of Blanco, who identified, among other plants, the species *Rhamnus lando* in a supplement to the *Flora*. But Llanos's entry on the species confused Merrill, who found that species identification and description did not match up. To Merrill, a Tagalog-savvy botanist would know that "lando" refers to a climbing shrub that fit Llanos's original description but not the identification.[25] Merrill corrected Llanos, reassigning the species under *Embelia*, a genus of climbing shrub, instead of *Rhamnus*, a genus of buckthorns, which he found more appropriate. Rigorous revision, therefore, demanded fieldwork and knowledgeable collectors to cumulate specimens *and* local names.

The Philippines and Clemens

Imperial combat at the turn of the century directed Clemens's career overseas. Before the outbreak of the Spanish–American War, the New York native—then Mary Knapp Strong—had studied the natural sciences at Williamsport Dickinson Seminary, a Methodist college in central Pennsylvania. Her captivation with nature began at a young age as she engaged in self-directed collecting work. While at Dickinson, she met Joseph Clemens (1862–1936), an aspiring Methodist pastor. Sharing religious conviction and intellectual passions, they married in 1896. The United States' war with Spain in 1898 opened a pathway for her first colonial plant-collecting trip through Cuba.[26] The couple bunked among US soldiers in Puerto Principe, and as Clemens collected, Joseph assumed a chaplaincy role for the military station. He applied for permanent chaplaincy and was assigned to the Philippines, where the couple first arrived in Samar in 1902.[27]

Samar had witnessed the bloodiest massacre during the Philippine–American War. Following a surprise attack on US soldiers in the town of Balangiga, US General Jacob H. Smith (1840–1918) ordered an attack that turned into the most widespread slaughter during the war years. One especially heinous atrocity was the torture and execution of eleven local guides, who had allegedly plotted to keep knowledge of edible plants from starving US soldiers during a march through central Samar.[28] Clemens's collecting work did not abate within this atmosphere of war, even as US military officers found themselves nervous about retaliatory violence in the couple's presence.[29] On the island, both she and Joseph acquired crafts, ritual objects, and natural history material, most notably avian specimens bound for US repositories, but the couple's 1905 meeting with Merrill and Edwin Copeland at the Bureau of Science after their Samar trip stimulated Clemens's curiosity for plant collecting.[30]

Armed with a plant press during their second trip to the Philippines in 1905, the couple boarded at Camp Keithley, a US military outpost at Lake Lanao on Mindanao. During their two-year stint, she collected over 1,200 specimens. Little collecting had previously occurred there, so Clemens's botanizing around Lake Lanao, which required a personal escort, earned her congratulations from US colonial scientists.[31] She had not made a habit of collecting local names of plants then and likely had few ready informants in the field to assist her with the task. Over time, however, her work demanded it.

Clemens collected across the United States before the couple moved their careers permanently to the Philippines. After serving in World War I, Joseph accepted a chaplaincy in the northern Chinese province of Chihli (or Zhili), and Clemens joined him to collect plants in the province. Joseph's superiors, however, discharged him back to the archipelago. Of the several reprimands that led to his discharge, including Joseph's own neglect of duties in order to botanize, the most serious was his refusal to identify a military officer who confided in him that the "Army didn't care about the morals of its men—only in their becoming fighting machines."[32] The couple bid farewell to their Methodist community in China and in the United States and resettled in Manila in 1922. The Bureau of Science served as Clemens's research base from their resettlement through 1936. The couple used the bureau's herbarium and library liberally. She worked unsalaried, fulfilling collecting contracts for US and European patrons as opportunities arose. Personal funds sustained the two until Joseph eventually became a conference evangelist, providing them with a regular income. Seasoned collectors like Ramos and Gregorio Edaño traveled with her in the Philippines and modeled plant collecting. "Treading in their steps," besides the "intoxicating thrills of glorious scenery and air" in the Philippines, she once confided, convinced her that botany was the only meaningful job.[33]

Ramos and Edaño not only identified choice material for herbarium accessions. They also had facility to negotiate with local residents and bureaucrats and to obtain local plant names. Metropolitan botanists were not consistently invested in the collection of local nomenclature, even though some colonial plant collectors in the nineteenth century defended their scientific import.[34] Clemens's earliest collections were not published with much local knowledge, including her first sets from the Philippines, but with time in the field and at the request of patrons such as Merrill, she recorded local details more meticulously in collecting notebooks and on labels she or Joseph affixed to sheets of mounted plant specimens. Unstudied in many of the languages that surrounded her during her fieldwork, Clemens relied on guides, translators, her Methodist missionary network, and "replying in mime" to collect as much information as possible.[35]

Clemens remarked how her "amateurish ear-marks" on her specimen packets required the pardon of her botanist-patrons. Humility—and self-deprecation— characterize her correspondence from this time. When writing to botanists, several of whom proffered her collections to major herbaria, she accented her amateurism, but with candor: "[H]aving no scientific reputation always makes me perfectly free to disclose my ignorance."[36] Like that of botanical artists, collectors, or female naturalist-explorers corresponding with gentlemen-scientists, Clemens's deference spoke more to social expectation and professional strategy than to her sense of her own scientific aptitude.[37] With select colleagues, she was not diffident but rather explicit about the shortcomings of other colleagues in the field. She bemoaned the decades of fieldwork she conducted in service to Merrill when remuneration for her labor—in her words, "labor gratis"—was inconsistent.[38] By the late 1920s and for the remainder of her career in the Philippines, she could boast botanical specimens obtained in China, British North Borneo, Java, French Indochina, and Sarawak. Her professions of humility were wise, given her professional standing and the social environment of colonial botany. She was well aware of her access to terrain otherwise inaccessible to stationary herbarium researchers working with dead plant matter.[39] Her immersion in the field (see figure 6.1), especially to amass material and knowledge, underscored her value.

Collectors Bound for Indochina

Years into their professional relationship, Merrill commissioned Clemens and her husband for an expedition to Indochina. By then, he had already produced a 1919 draft revision to *Flora cochinchinensis*, but he had evidently hit a wall. Dozens of Loureiro's identifications were too murky without physical material and more local data. In May 1927, Clemens and Joseph departed Manila for Huế (or "Hue," as it appears in historical correspondence) on Indochina's central

FIGURE 6.1. Mary Strong Clemens, c. 1930s, holding nos. 1744–1755, Bartlett-Clemens 1930–1939, box 5, correspondence and associated materials, 1875–1959, University Herbarium, Bentley Historical Library, University of Michigan. Reproduction permission courtesy of the Bentley Historical Library, University of Michigan.

coast.⁴⁰ For the couple, Merrill secured grants from the New York Botanical Garden and a private San Francisco–based donor.⁴¹ They were to complete the trip under the auspices of the University of California, where Merrill had taken a deanship at the College of Agriculture in 1922. In return, he stressed that "specimens from Hue with notes and local names [were] absolutely essential."⁴² Ahead of the trip, Clemens had little time to prepare her travel documents, which included passport photos and letters of introduction to facilitate her travel through the French colony. US colonial government personnel endorsed her letters, one of which would initiate contact with a veterinarian-botanist near Huế who could assist Clemens with plant names.⁴³

Loureiro had spent decades surveying the natural history of Cochinchina and China in the mid-eighteenth century. Like his Augustinian Philippine counterpart, Blanco, he had grown curious about materia medica.⁴⁴ Originally deployed in 1742 to undertake missionary activities, he eventually served as a naturalist for the Nguyễn lords and resided in Huế.⁴⁵ He completed the manuscript for *Flora cochinchinensis* in 1788, and two years later, the Academia das Ciências de Lisboa published the 744-page volume. Though sweeping for its time, the flora posed several taxonomic problems in the twentieth century for Merrill.

To approach the revision, Merrill looked to Clemens to gather material that could match Loureiro's eighteenth-century enumerations largely based on the flora of Tourane in central Indochina. He also needed to compare Loureiro's plant binomials to those proposed by other botanists who had worked from *Flora cochinchinensis* since its publication. Finally, local names, paired with any known uses, aided more precise identification. Merrill completed *An Interpretation of Rumphius's Herbarium Amboinense* (1917) and *Species Blancoanae* (1918) with the same combination of methods: a review of the literature and field research. On local names, Merrill explained, "While [they] must be used with caution, and while the specimens bearing them must of necessity be critically compared with the original descriptions of the species it is suspected the specimen may represent, still in a very high percentage of cases the local name will give the clue to a large number of imperfectly described species of the early authors when all other attempts to locate them have failed."⁴⁶

In his manuscript, he bemoaned the absence of local nomenclature in herbaria. Such data, he believed, could be of use to the modern systematist. Some earlier botanists were of the same persuasion, such as Michel Adanson, who espoused "a pragmatic populist approach to naming" that refused Latin as the sole language of the science, and Jean Baptiste Christophore Aublet (1720–1778), whose contemporaries dismissed as "vulgar" the generic names for French Guianese flora he had drawn from local nomenclature.⁴⁷ William

Colenso (1811–1899) valued Māori nomenclature, finding indigenous plant knowledge and philological utility in it.[48] Miguel Colmeiro, writing in 1885, insisted that common names "must be known to scientists" because etymologies "contain useful and instructive memories."[49]

In Merrill's case, he expressed the mutability of Latin binomials and the *im*mutability of local plant names: "We pride ourselves on the assumption that the Latin binomial or technical name of a species is theoretically fixed, but in this case theory and fact are not in agreement, for, due to one cause or another, changes in binomials are exceedingly frequent in modern taxonomic work."[50] Synonymy, to Merrill, had become a problem in the science. Before him, Candolle portrayed synonymy as an "evil" that emerged from "very numerous," "partly inevitable" causes.[51] Perhaps most troubling was the aggregative nature of synonyms. A botanist unaware of a plant's previous names and descriptions could give the plant another Latin name, prior publications notwithstanding. Moreover, what was considered a valid venue for publication varied through the early twentieth century. As botany internationalized, many botanists, including Merrill, believed the tapering of such venues to academic monographs or serials meant efficiency for the discipline.[52] For others, it could mean gatekeeping and the foreclosure of the science in service to professionalization.

Merrill declared that the "local colloquial name of a species," especially one with economic importance, had a stability that spanned centuries and would "continue to be used for centuries to come."[53] The Philippines inspired Merrill's stance. Without voluminous Spanish writing on the archipelago, he recognized that Latin alone could not be the unifying lingua franca across the Spanish and US colonial regimes. Instead, Philippine languages offered a continuity of knowledge that, for colonial botany, dropped precipitously after the destruction of Manila at the outbreak of revolution. Held among those fluent in various languages and dialects, plant knowledge circulated outside and in spite of Latin's mediation. In both manuscript versions of his "Commentary," he cited examples of Philippine plant names first referenced in his 1903 *A Dictionary of the Plant Names of the Philippine Islands* and their etymologies that, he presumed, pointed to the persistence of certain local nomenclature. Although dating the philological emergence of non-Latin names can be challenging, Mercedes G. Planta has argued that one may presently find the durable character of some plants' local names even with the archipelago's linguistic variety.[54]

Though Merrill reiterated the importance of local nomenclature in his 1934 manuscript and the finalized publication in 1935, his own editorial interventions belied a still unsettled perspective.[55] In the 1934 manuscript, he nixed his statements on the "very definitely fixed" name for species of local importance

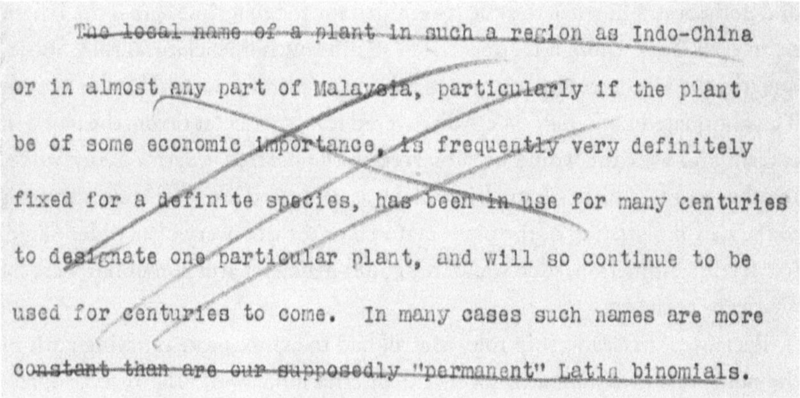
The local name of a plant in such a region as Indo-China
or in almost any part of Malaysia, particularly if the plant
be of some economic importance, is frequently very definitely
fixed for a definite species, has been in use for many centuries
to designate one particular plant, and will so continue to be
used for centuries to come. In many cases such names are more
constant than are our supposedly "permanent" Latin binomials.

FIGURE 6.2. Detail of Elmer D. Merrill, "A Commentary on Loureiro's Flora Cochinchinensis," 137, manuscript, 1934, ser. 3, Elmer Drew Merrill Papers, Archives of the New York Botanical Garden. Reproduction permission courtesy of the Archives of the New York Botanical Garden.

and their fixity for centuries to come. Deliberate pencil marks strike the line, "constant than our supposedly 'permanent' Latin binomials" (see figure 6.2).[56]

This sentence does not appear in the 1919 manuscript, though he added it to the 1934 manuscript only for it to be removed before publication. The deletion could have signaled his change of heart on the matter. Instead, I suggest the scrapped statement indicates the wider intellectual politics of the time, reflecting major developments in international botany in the first third of the twentieth century. While Merrill bristled at the intellectual current that assumed Latin's faultlessness, he published in congruence with it as his influence in the IBC grew.

Revising the IBC Nomenclatural Standards

At the 1930 IBC assembly in England, Merrill served as the vice president and chairman of the Section on Nomenclature. Proceedings of the IBC noted that he "very ably" led the section, despite having "undoubtedly the most difficult position in the entire congress."[57] Up to that point, the IBC had witnessed and incited some of the most bitter nomenclatural debates among opposing imperial schools. In 1905, the IBC ratified the "International Rules of Botanical Nomenclature," which required a Latin description of an accepted taxon and established lists of *nomina generica conservanda*, Latin names protected against any renaming. From this meeting, a dissenting group of US botanists formed the "American code" that, among other stipulations, opposed Latin as the language of description.[58] In 1926, the IBC assembled for the first time across the Atlantic in New York, though it adopted no significant nomenclatural rule. Subsequently, the 1930 meeting reaffirmed the "principle of priority." Under it, even if an antiquated Latin name were discovered for a particular taxon, the name in current and wide use would be preserved in the *nomina generica conservanda*. In other words, a plant's Latin binomial was protected even if disputes arose as to the new designation of the plant material or the discovery of an older name. In Merrill's estimation, such standards could manage Latin names but were by no means perfect.

Because of his leadership role, Merrill had to evince more unifying faith in the nomenclatural standards set by the international body. The 1930 congress remains noteworthy in the history of international taxonomy because competing schools merged conventions that produced a more trans-imperial nomenclatural system. For example, the American code–endorsed "type concept"—that is, the selection of a physical specimen to represent a taxon—became part of international practice. The type concept proved critical to the 1935 "Commentary" because Clemens's collected plants were eventually acknowledged as type material for several of Loureiro's species. Although Loureiro had collected some plant material, little survived by the time Merrill set about revising *Flora cochinchinensis*. The IBC's concession to aspects of the American code was not trivial and likely impacted Merrill's own diplomacy. His 1934 statement was then, perhaps, hastily written, revealing his displeasure with the loftiness of Latin binomials, their users, and the IBC's long-running debates. Nevertheless, Merrill conceded that "Commentary" had been "consummated under the general provisions of the International Code of Botanical Nomenclature." The work also followed Engler–Prantl systematics, then agreed to be the most up-to-date system of arrangement. He accepted nomenclatural directives even when his personal opinion may have diverged.[59]

A Renewed Sovereign

With the approach of the sixth IBC in Amsterdam in 1935, Dutch botanist Marius Jacob Sirks (1889–1966) wrote of the need for "international cooperation," remarking that the previous meetings demonstrated "a strong feeling for international affinities."[60] "International" in the age of new imperial science was, of course, overbold.[61] In 1934, with the exception of Japan and South Africa, European member nations comprised the Union of Biological Sciences and its Botanical Section.[62] All meetings through 1935 had been in imperial metropolitan cities.[63]

For the 1935 assembly, Merrill served as president of the Subsection on Nomenclature.[64] He idealized international botanical cooperation. In the field of taxonomy, he campaigned for, among other improvements, "the durability of the formation" of an international body to oversee plant taxonomy; the end to research duplication and specialization in order to dedicate more study to lesser-known plant groups; and "the development of inter-institutional cooperation in the botanical exploration of those parts of the world as yet inadequately known from a botanical standpoint."[65]

By the mid-1930s, Merrill was not the only botanist advocating such cooperation. Dutch bryologist Frans Verdoorn (1906–1984), also a plant collector of Malesia and the Dutch East Indies, expressed an internationalist call. Verdoorn served as editor of *Chronica Botanica*, an internationally oriented botany serial. Debuting in 1935, the publication boasted reports in the imperial languages of English, German, French, Spanish, and Italian.[66] In 1934, Verdoorn invited Merrill to the journal's advisory board because he had heard of Merrill's commitment to work of an international bent.[67] For the maiden issue, he asked Merrill to write the publication's leader on the topic, expressing how taxonomy, "the oldest branch of botany," showed "better than any of the other branches the need of international collaboration." Merrill had been delayed in reaching Verdoorn about the leader—a delay that Verdoorn suspected had resulted from the politically sensitive nature of the topic given the "present conditions" that showed "a victory of the doctrine of nationalisation."[68]

Concerns over the rise of Nazi Germany waft through international botany discourse from this time. Botanists like Merrill did not mount explicit critiques at the country's violent totalitarianism. He instead framed Germany as an uncooperative botanical community that had made the need for international cooperation more urgent. German participation in the International Union of Botanical Sciences was allegedly patchy, and German scientists had been excluded for some time during deliberations.[69] In 1935, Merrill commented about the prohibitively high cost of German technical periodicals, which after the Great Depression amounted to "extortion." According to him, German periodicals sold for five to eight times the rate of those sold elsewhere. He notified his colleagues that he would cancel sixteen German subscriptions and directed other US institutions to do the same to reduce the cost.[70] Verdoorn's ideals and Merrill's call to boycott German publications suggest an internationalism that had been building in the sciences since the second half of the nineteenth century. Amid the budding and hyperviolent manifestations of nationalism, their work is arguably part of the "international decade" of the 1930s.[71] As international collaboration occurred on a practical level—among botanists, collectors, field assistants, and illustrators—it emerged rhetorically to stem nationalism's ills, as well.

After a unifying victory for nomenclatural standards at the fifth convening, Merrill chose to implement the new IBC guidelines at his home institution, the New York Botanical Garden. He started his directorship of the garden in 1930, arriving to a facility and staff wrung by the Great Depression.[72] By his second year on the job, he contemplated leaving, especially because he lacked support from staff and the garden's board of directors. One board member in particular vehemently disapproved of Merrill's views on nomenclature. The dissent and displeasure with his leadership, Merrill believed, stemmed from his implementation of the IBC code and the elimination of the then-defunct American code.[73]

Instituting the international standard at the expense of his own popularity at the garden was a weighty choice. The Depression had severely limited his budget, oversight, and ability to carry out original research, even with his Philippine-born colleagues and the former collectors with whom he had worked.[74] To defend his decision, an earnest concession to the international agreement that had been years in the making, he removed traces of his own cantankerous stance. In this institutional and political environment, *Transactions of the American Philosophical Society* published Merrill's 444-page "Commentary" in 1935.

Collecting in Indochina

In the final "Commentary," Merrill communicated his indebtedness to Clemens's botanical collecting. That acknowledgment did not prevent him, however, from openly lamenting that had all the local names been consistently recorded, even the plants with doubtful identities would have been resolved.[75] She and Joseph had marshaled as many local plant names as possible, but the job proved challenging. Rain inundated their camp. Leeches overcame them. Ants ventured to devour their collections. Inexperienced with French or Annamese, the couple could only uncover so much local nomenclature. Joseph admitted, "[Mary] is labeling, but not much doing with Annamese names. The old French medico who is studying plants of his profession, declares it to be a meaningless thing, as each village has a different name and no one is wise on plants."[76]

Even so, the couple maintained a strange vigor: Clemens "lives on botany," her husband said admiringly. Luxuries and local expertise made their work easier. The couple enjoyed a rented automobile and special clearances from the local mayor, who equipped them with a permit "to collect anything."[77] At least two remunerated Annamese field assistants collaborated with them during the first half of the collecting trip.[78] The assistants not only dried specimens in the field but also brought back unique material after a day's work. From his six-kilometer return home, one collected and offered six specimens unfamiliar to the couple.[79] However, the couple's doggedness exhausted the assistants, who became so seriously unwell that one was transported to a nearby hospital.[80] Aiming to maintain their pace, Clemens and Joseph hired another three young men, who contested their unrelenting work. In the middle of a day's excursion, one sought respite at home during the group's usual mealtime. He opposed the couple's refusal. Joseph proceeded to discipline him, and one can deduce that this included physical force: "a few raps on the head," he claimed, "are helping a lot."[81]

With their assistants in Indochina, the couple covered the vicinity of the Y Pha Nho colonial cemetery, An Bang cemetery, and likely Sơn Trà mountain. By the second half of their trip, they no longer referred to their first two assistants in their correspondence, and what became of them and the three other young men remains uncertain. The couple continued to rely on field labor, this time recruiting two remunerated ethnic Chinese men.[82] At least one of the Clemens's mounted specimens (see figure 6.3) suggests these men were also actively collecting material on their own, aware of plants the couple overlooked. The sheet for the leaf vegetable *Sauropus androgynus* mentions an assistant's collecting work, reading "Brought by coolie," using the now derogatory reference to Chinese labor.

Present in local dishes, the popular green could have been familiar to the assistants. The surviving original herbarium accession features stems, petioles, leaves, and fruits, which show the assistant's collecting aptitude. Collecting fruiting structures is necessary for accurate identification and adherence to the standards of an herbarium accession. Clemens's patrons ridiculed her for collecting partial or insufficient samples that were likened to scraps or detritus. The German mycologist Paul Sydow (1851–1925) was notoriously insulting about Clemens's work and at times grumbled about her collections to her superiors at the Bureau of Science.[83] Merrill admonished her for collections completed after Indochina, critiquing her fragmentary field notes, disorganized sets, and dried specimens that in some cases were "mere scraps . . . worthless to any institution."[84] This derision, however, was commonly lobbed at most self-taught collectors as botany professionalized. US botanists gradually withdrew from their association with amateur "botanizers" as collaboration yielded less social and financial capital.[85] While this rift emerged in the metropole, amateurs were still vital to the understaffed US colonial botany operations in the Philippines. Botanists could fling critiques at amateurs, who were often locally born or female,

FIGURE 6.3. *Sauropus androgynus* (L.) Merr., Annam, Tourane, J. Clemens with M. S. Clemens. Muséum national d'Histoire naturelle (MNHN), Paris, Collection: Vascular plants (P), Specimen P00318266. Clemens's specimen label reads, "Glochidion. Frt. white. Brought by coolie. Brob. thicket by dunes." The field assistant likely collected the specimen on June 17, 1927, which was then processed as an accession in the MNHN herbarium on May 8, 1928. Today, the plant is referred to by some northern communities as rau ngót (literally, "leaf that contracts" upon cooking) and as bù ngót and hay rau tuốt ("stripped leaves" during culinary preparation) elsewhere. Reproduction permission courtesy of the Muséum national d'Histoire naturelle, Paris, Herbarium (P). The MNHN gives access to the collection under the RECOLNAT national research infrastructure.

but their professional prestige rested on the amateurs' ability to navigate lands they could not comprehensively survey on their own.

At their trip's end, the couple arranged and shipped to Merrill ten boxes of 1,200 specimens, several with duplicate sheets to be sold to various herbaria.[86] Joseph had taken the sheets to a French assistant forester, who labeled each package with an Annamese local name. Joseph advised Merrill, "Perhaps you can judge the worth of his [labeling] work, but sprinkle it with salt."[87]

Ordering Babble with a Vernacular

Clemens felt Merrill discounted her labeling. Her labels, however, were not entirely rubbish. In fact, Merrill extensively used her annotations and the information her assistants and French interlocutors provided. Loureiro, he discovered, described the same species as different genera, at times unaware that the number of stamens and carpels of a species might vary from sample to sample. Loureiro also separated the flowering specimen and the fruiting specimen of the same taxon into different genera. The structural mix-ups included Loureiro's *Nerium antidysentericum*, a dogbane species. Clemens collected a sample she generally identified as *Wrightia*, which in her view sported red flowers.[88] Her specimen matched two different taxa descriptions from *Flora cochinchinensis*: *N. divaricatum* in general character and the corolla (or petals) of *W. annamensis*, whereas Loureiro suggested the *W. annamensis*'s flowers were white-green. Merrill agreed with Clemens's visual judgment and cursory identification, concluding Loureiro had mistakenly claimed the original as a new species, *N. antidysentericum*.[89] He identified Clemens's sample as *W. annamensis* and corrected the work to reflect that.

Merrill also compared the local names she obtained with those that Loureiro and other botanists had recorded. A Portuguese missionary developed a Romanized script for the prevailing local language in the mid-seventeenth century, but it had not been in compulsory use until the French colonial government implemented it in 1910. For the catalogs containing local names, naturalists and botanists attempted to follow the Roman script but did not implement a standardized system of transliteration. Nevertheless, cognates, as Merrill referred to them, assisted his work.

Merrill identified cognate resonance for species like cây tam buot (as written in "Commentary"), a sour fruit resembling a gooseberry. In Merrill's opinion, Loureiro had likely described a species similar to one described by a later French botanist, Lucien Beille (1862–1946), who also published on Indochinese flora. Beille, who had not consulted Loureiro's work, accounted for a similar species under the Latin name *Phyllanthus distichus*, for which he included the name "cây tam mot." To Merrill, the cognate relationship between the two local names was unmistakable, with "buot" and "mot" bearing similar pronunciation along different transliteral rules. When read alongside Beille's description and Loureiro's, as well as another nineteenth-century commentary that suggested that the species was identical to *Cicca acidissima*, Merrill had confidence he had identified the specimen correctly. *C. acidissima* had become a synonym of *C. acida* by then, so Merrill identified cây tam buot as *C. acida* in "Commentary."

Likewise, cây bap benh, a common medicinal, had previously been cataloged as ba binh by another French botanist, Paul Henri Lecomte (1856–1934). Loureiro identified it as *Crassula pinnata*, but the description and the local name instead matched succeeding descriptions of *Eurycoma longifolia*, with which Merrill agreed. Merrill applied a similar logic to co mè, an edible weed with medicinal uses. French botanist Charles-Joseph Marie Pitard's (1873–1927) logging of coc ma led Merrill to suspect that Loureiro's earlier mixed identification of *Pharnaceum incanum* actually referred to *Oldenlandia corymbosa*.[90]

Botany's Entextualizing Authority

Local nomenclature acted as a sonically distinguishable hum through an otherwise babbling crowd. Merrill highlighted twenty-three species that would have otherwise been indeterminable without the non-Latin names, relying heavily on Anglo-European source material that had recorded such names. His reliance speaks to a particular authority inscribed in the recording of the nomenclatural vernacular.

Even as Anglo-European botanists displayed polyglot tastes, they engraved their own authority when publishing local nomenclature—nomenclature entextualized during field research. In the colonial context, the process historically detached speech from the conditions of its utterance to often fix it as "'authentic' cultural text." This process happened in nineteenth-century colonial India, for example, where administrators cataloged folklore and proverbs shared by Indian informants and reported on them as though they were a "monologic," consensual view reflective of all locals' perceptions of caste.[91] To achieve a degree of consensus, the compilers of the eighteenth-century *Vocabulario de la lengua tagala*, Juan de Noceda and Pedro de Sanlucar, claimed that the accent, meaning, and pronunciation of each word had to be agreed upon by twelve natives before inclusion in their dictionary.[92] Colonial Anglo-European botanists' and collectors' entextualizing practices echoed this tendency to fix meaning and language.

The "fixity" of local plant names could prove useful, yet the social conditions under which such information was obtained are typically absent, begging the questions: How were local names collected? Who uttered them? Who asked? In whose company and under what social agreements? References to or indices of local names imply nomenclatural harmony across their users, effacing a multivocal history of contestation, consensus, and deletions.[93]

Collecting labels that attempt to standardize the knowledge obtained during field research may mask the particularities in the field, not least of which includes the political atmosphere within which a colonial-foreign collector or botanist worked.[94] A return to Clemens and Joseph's correspondence reveals the social conditions that demonstrate aspects of colonial entextualization. Merrill's finalized "Commentary," rather than gesture to those conditions, presents a species such as *C. acida* next to cây tam buot as fixed information, empty of the political entanglements that contributed to the authoritative inscription. Those included France's colonization of Indochina (1887–1954) that predicated the arrival of French personnel and intellectuals who grew familiar with local flora. That a French "medico" and forester were at the Clemenses' disposal reflects, as well, the global imperial field that emerged at the turn of the century. The olden days of botanical imperial competition—marked by clandestine operations and "colonial espionage"—bore a new (but not altogether different) cloak, one hemmed with an internationalist thread.[95] Intercolonial networks thus emerged to further more imperial consolidation. Plant presses, labels, and specimen papers were not the only important components of the couple's arsenal—so were letters of introduction written by US colonial personnel that could ensure safe, prolific work in a French colony.

These political dynamics both enabled and hindered Clemens's success during her botanical research. That Clemens and her husband could collect plant specimens early in their career under the protection of US military illustrates the entwining of colonial botany and military occupation, as explored in chapter 4. Merrill's original deployment to Manila has similar overtones. As a chaplain, Joseph had been armed with a revolver, as was the case during the couple's first deployment to Samar. Months following the Balangiga massacre, he journaled about armed US soldiers haphazardly shooting in the dark of night, paranoid about a local assault on their camp.[96] The very outposts where the couple was stationed in the Philippines sought to defeat local insurgents, whose historical racialization amplified military violence.[97]

The couple's position as Methodist missionaries, particularly among the Moro population of Mindanao, also privileged them with the backing they received from the US military. The sprawling Protestant network that partly undergirded US colonization of the Philippines helped Clemens and Joseph's travels. Their Methodist colleagues ensured their comfortable passage throughout Indomalaya.[98]

They diligently recorded the number of "souls saved" on their travels: preaching, praying, and missionary work were foundational to their careers. To a fellow missionary, Clemens once effused, "Botany is a long, long, road but having put your hand to the plow you must press onward forever and more and more you will worship and adore the marvelous Creator of all things beautiful, O, the depth of riches of the wisdom and knowledge of our God!"[99]

Clemens's position as a self-employed white female collector may have garnered harsh judgments from Anglo-European male botanists, but her identities likely stimulated connections in the field otherwise closed off to her dismissive male colleagues. She holidayed with Ramos's wife, was privy to the personal details of the bureau's Filipino and US personnel, and maintained warm relationships well after her permanent departure from the Philippines in 1936. Clemens's independent collecting—with mixed funding, institutional freedoms, and mobility—also becomes more vivid when compared to Ramos's privileges as a salaried Filipino bureau employee. Based on the specimen sales from her three-month expedition to Indochina, Clemens earned nearly five times what Ramos earned in one month of employment at the Bureau of Science at the height of his career. Furthermore, Clemens's network and white foreigner status ensured her freedom of movement regionally, while Ramos was confined to the Philippines and only once participated in a collecting trip to Borneo. Both were considered amateurs at the start, yet their work contributed to Merrill's own published record to cohere Indomalayan regional and Philippine proto-national flora.

Colonial entextualization also obscures the Annamese and Chinese assistants behind Merrill's "Commentary." The conditions of their labor fade behind the local names and even the labels that remain fixed to herbarium sheets. A reexamination of the labels on the *S. androgynus* accession demands pause. "Brought by coolie" points to who assisted the couple on their Indochina expedition. Yet it overlooks why the couple hired Chinese laborers as well as Annamese assistants. It further hides the Clemenses' relentlessness in the field, and their "hope" they would not "kill off too many of the natives who [tried] to take the pace."[100] Indices or references to local names rarely reveal the controversy, arbitration, force, collaboration, or social maneuvering that occur during field research, let alone the accidents, illnesses, and inclement environments that make the work daunting. Local names in "Commentary" beside the archival record encourage contemplation of what the field assistants gained and the knowledge that they bore.

That the Clemenses' field assistants were paid and accompanied guests of the colonial government could convey a certain amount of social capital. They independently retrieved specimens on the Clemenses' behalf, demonstrating their cogent familiarity with the local environment. As Clemens herself endeavored to paint the best image of flora in a given locality, doing so usually entailed collecting within the geographical boundaries of a particular location. To an untrained observer, distinguishing all species in a locality can be tedious, but for the Chinese assistants on the couple's Indochina trip, acquaintance with local plant life allowed them to spot growth unbedecked by bright flowers or conspicuous fruits. *S. androgynus* was not included in the finalized "Commentary." Loureiro likely did not reference the species or anything similar to it in *Flora cochinchinensis*. Its inclusion in the couple's study set nevertheless reveals the sifting through of local knowledge that botanists could value—or simply overlook.

Reexamination of the local names presented in "Commentary" shows, on the one hand, the isolation of speech from a particular context and on the other, how utterances captured in situ or recalled by other informants become "recontextualized" within the field of international botany. Thus, local names become the only and most authentically long-lasting bits of information. And yet, the original cognates that appeared in Merrill's revision depart from today's usages and nomenclatures. For the gooseberry, present-day communities in southern Vietnam refer to the plant as chùm ruột, different in presumed pronunciation from previous records. "Ba benh," the popular medicinal of Merrill's work, can be understood in relation to today's bách bệnh among northern Vietnamese speakers or the "bá bệnh" used elsewhere: the former literally translates to a plant used to remedy "one hundred ailments," while the other refers to remedying "many"— meanings that fade behind Merrill's systematic work while showing dynamism and variability. The history of inconsistent transliteration in French colonial materials notwithstanding, a certain impermanence should be expected given the irregularities of transliteral forms, spoken usage, and recording.[101] Nonetheless, local names continue to exist in botanical publications as though local informants and collector-botanists share the same intellectual agenda.[102] The history of "Commentary" shows that this consensus is much messier to determine.

* * *

Clemens and Merrill's collaboration offers another take on synonymy, which I consider a euphemism for botany's Latin nomenclatural disorder. During the first third of the twentieth century, international botanists worked against the Latin babble to institute lists, publications, and type-specimen standards that could reduce the freewheeling application of Latin names. As Merrill attempted to correct *Flora cochinchinensis* with a vernacular, he used the long-dismissed Babel to correct the babble. As I have shown, the assumption that local names are more fixed than Latin raises questions not easily answered when casually reviewing a list of names associated with a single species. Moreover, the assumption gestures toward the very desire "to fix" names for certainty—Latin and non-Latin.

Whereas botanists could posit Latin binomials as a universal system to delineate species, assumptions of the "very fixed" nature of local names operate within the same rubric, albeit on a more localized scale. Much of the history of imperial botany concerns the mobilization of different knowledge for universalizing ends, a generally understood ambition of Enlightenment-era science and its successors. What becomes even more provocative is how fixed local names could be mobilized in that effort. Yet, the colonial history of their acquisition reiterates the process, neither neat nor straightforward, by which names were obtained. What's more, such presumptions of nomenclatural fixity perpetuate characterizations of indigenous communities and colonized peoples as static, if not primordial. The stasis trope suppresses the manner in which other ways of knowing are historically made and unmade. To reserve debate, change, and evolution for a particular intellectual community—and not just among botanists—denies them to those whose knowledge-claims and knowledge-making are deemed "timeless."[103]

Conclusion

OF PLACE, MOMENT, AND SOURCE

The Rose of Alexandria

(*Rosaceae family*)

This is one of the flowers that helps beautify a garden immensely. This may be due to the pleasant scent of the petals or owing to the fair color of its "varieties." But in truth, its propagation assumes a large place, as many are of the opinion, such that a garden without roses is like a home without a maiden: one without cause for joy, comfort, and... hope." — "CONCERNING FLOWERS," *Renacimiento Filipino*, Tagalog Section, 1911

For a time, the trilingual serial *Renacimiento Filipino* ran a Tagalog column titled "Concerning Flowers." In 1911, the paper published a gardening guide for roses, the woody perennial suspected to have originated from Central Asia. The guide opens with an appeal to love, the satisfaction of "being the one who chooses and the chosen object" in the same way a young woman might pick a fragrant flower for herself. The author references the Rosaceae family, to which the rose is understood to belong, and is convinced that the Alexandria variety must have come from Egypt given its name. Despite its faraway provenance, the potted rose can withstand the sun's heat in a place like the Philippines. Watering before the "sun peeks" and again "before it retires for bed" can ensure a resilient plant. Handfuls of horse dung and mud fertilize rose roots well, though should be used sparingly. The water left after rinsing raw fish, on the other hand, can be used every other day, and if a fern-like parasitic fungus invades the soil, a little sprinkling of sulfur should do the trick.[1]

"Concerning Flowers" captures much of what has been discussed in previous chapters. The popular column written in a language of the Philippines combines elements of Linnaean botany, matters of the "variety," the gendered axes of botanical life, a consideration for place, and knowledge of a plant uncommon in the conventional colonial flora of the time. At its core, the column speaks to the sovereign vernacular, at once contributing to and yet departing from botany's strictures.

This book has provided a history of botany in the colonial Philippines. The notion of the islands as one cohesive space invited analysis by foreigners and locals alike. Dozens of Spanish and US botany personnel set sail to account for every single plant specimen across the islands. Foreign botanists, missionaries, and local collaborators of many stripes worked to make Linnaeus's dream real. The logic of botany—that all vegetal life on the planet can be recorded and systematically arranged to show relation—made their impractical objective seem achievable. Spanish and US botanies shared this aim and the grammar of a new kind of imperial collaboration that marked a significant time of colonial transition in the Philippines. The final and first decades of Spanish and US colonialism saw the modern proliferation of the science. Imperial botany took on new color as an international science, uniting botanists representing their national flags as one body. I close this book by comparing the history of botany under the Spanish and US empires in the Philippines, by returning to what such a comparison does for the historiographical recuperation of local actors, and by considering the methodological potential of the sovereign vernacular.

A More Symmetrical History

One of the objectives of this book has been to detail botany as it was practiced under the Spanish and US empires in the Philippines. Doing so has permitted an examination that departs from accounts of science in the colony that see the two regimes as overwhelmingly different. The year 1898 was inarguably a consequential one in Philippine history, but turning it into a chasm prevents a clear understanding of the local and global histories of botany and the many continuities that spanned both eras for the archipelago.

Differences existed between the two. Institutionally speaking, their hiring practices were distinctive in order to address labor shortages that each regime encountered when erecting botany institutions. The Spanish hired in a narrow manner: officials came either from Spain or the Philippines. Dozens of men applied for low- and mid-ranking service, aiming for a coveted government post. Even if institutions distanced themselves from the once intellectually secretive practices of the Spanish monarchy—the *arcana imperii* of the Iberian Peninsula that kept imperial records unpublished—the Jardín Botánico de Manila (JBM) did not have other foreign nationals regularly walking through its doors, collecting botanical matter, or overseeing provincial lands.[2] The US Bureau of Science did. The impetus to invite, hire, and collaborate with non-US and non-Filipino researchers, as well as with self-funded and independently employed collectors, came from a very clear personnel need. Even with the College of Agriculture in Los Baños and a cadre of US-trained pensionadas and pensionados returning to the colony, many independent researchers brought their intellectual appetite for the natural sciences to the Philippines. Mary and Joseph Clemens were among them.

Comparatively, US colonial botany was a publishing machine. Elmer D. Merrill was not alone in his productivity. The Philippine colonial administration, the Bureau of Science, and foreign naturalists published regularly on Philippine plants. Works included agricultural treatises, commercial profiles of product-bearing flora, such as those on abacá, and descriptive botany. The *Philippine Journal of Science*, the longest-running science periodical in the archipelago today, gave US colonial scientists and those in "adjacent countries of the Orient" completing "scientific work of the proper character" a venue to share the latest in plant discoveries and experimentation.[3] Of the five research essays in the journal's first issue in 1906, three dealt with a local palm, the coconut. Even though US botanists upheld their expertise as paramount, this did not keep others such as Regino García from publishing on rice varieties, among other botany topics, under US colonial auspices.

Spanish operations, on the other hand, did not have as many publishing specialists. The Inspección General de Montes (IGM) and the Comisión de la Flora y Estadística Forestal de Filipinas produced the greatest number of publications seen in secular colonial Philippine botany under the Spanish. Still, only a few individuals so far known penned significant manuscripts: Sebastián Vidal, his brother Domingo, Ramón Jordana, García, and a small collection of others. Limited state coffers could explain some of this stunting. Institutional reports and published surveys comprised these individuals' main intellectual production. These appeared alongside the late nineteenth-century natural history writings of other foreigners, such as Fedor Jagor. The Spanish did not establish science periodicals to which foreigners could contribute, and foreign naturalists did not publish under the auspices of Spanish colonial scientific operations at the volume they did under those of the United States.

While acknowledging these and other differences between the two colonial botanies, this book has brought forward the institutional, intellectual, and political similarities between them. These similarities contribute to a growing Philippine historiography that has sought to complicate the colonial transition. First, both Spanish and US colonial botanists dedicated the infrastructure and personnel resources to develop botanical libraries, herbaria, and research gardens. Their herbaria preserved specimens from the archipelago and from neighboring colonies, revealing both Spanish and US interest in intercolonial plant exchange. Even if members of the public criticized the JBM's floundering, by the close of the nineteenth century researchers cultivated trees and plant species not endemic to Manila and surveyed islands outside Luzon, and several affiliated artists produced some of the most visually innovative and sumptuous botanical works seen in Philippine history.[4]

Second, both colonial administrations developed programs to train local men in land surveying, agriculture, botany, and forestry. In addition to its training program, the JBM's Escuela de Agricultura also provided a student-worker program that remunerated students as they completed their studies. The US colonial College of Agriculture provided theoretical and field preparation for men, some of whom went on to decades-long careers in botany. Both institutions' pensionado-style method of recruitment of local students suggests that their institutional practices were more similar than has been previously remembered. Long associated with the US colonial state, the pensionado program alluded to earlier funded a select number of local students to train academically in the US with the expectation that those students would return to serve the colonial administration in the Philippines. Spain instituted a similar program through the JBM. Though local students were not sent to the metropole, students from provinces distant from Manila received funding to study with the expectation that they would return to their home provinces and propagate modern botanical practice.[5] Filipinization of the bureaucracy may have been a US-era initiative, but both pensionado programs subscribed to an intellectual indoctrination to serve shared colonial visions, which included the botany-informed, commercially promising exploitation of the Philippine environment.

Third, both colonial botanies responded to major developments in international botany. Zoilo Espejo and García followed Candollian systematics to arrange the JBM's annual seed catalogs. With the advent of Bentham and Hooker systematics, Vidal arranged his co-produced *Sinopsis* following the new system. US botanists subscribed to Engler-Prantl systematics, as seen in Merrill's commentary on *Flora cochinchinensis*, which coincided with the rise of Darwinian evolutionary theory. These adoptions were not trivial. They evidence Spanish and US responsiveness to global intellectual currents and the Philippines' participation in them. During the political upheaval on the peninsula in the second half of the nineteenth century, Spanish botanists could not play a significant role in the direction of the International Botanical Congress (IBC). This did not mean they stayed isolated in their pursuits: scientific statecraft ensured that Spain could present itself as a relevant intellectual player on an imperial stage. US botanists wanted to surpass the institutional legacies of European botanical theory and practice. Doing so secured major US influence over the IBC in the first third of the twentieth century, even as Merrill truckled to the nomenclatural conventions of the governing body.

Finally, as people, publications, and plant specimens moved widely at the turn of the century, the possibilities for a more regionally oriented botany emerged. Prior to this period, naturalists working as early as the seventeenth and eighteenth centuries considered the idea of a European regional flora, comprising the efforts of Europe-based intellectuals, to identify a common geographical spread of certain plants.[6] Studies in plant geography, or the geographical distribution of species, took fuller form in the early nineteenth century, perhaps most commonly associated with the work of Alexander von Humboldt (1769–1859), whose own seminal treatise on Spanish America came from the empirical insights of local intellectuals.[7] "Regionalization" of plants and animals saw an uptick from the mid-nineteenth century onward.[8] By the end of the century, in the age of international politicking, foreign and local functionaries working in colonies drew up these regional connections in a more pronounced way.

Both botanies took to the work of constructing regions. Malesia became the regional framework within which Spanish, German, and Italian botanists shared their intellectual production. Vidal immersed himself in interimperial intellectual exchange by engaging with German botanical material and vice versa: German botanists came to rely on Vidal's work while developing botanical studies of what had been forming as a floristic region. Writing over sixty years after the Spanish state's acquisition of Odoardo Beccari's *Malesia*, Cornelis G. G. Jan van Steenis (1901–1986) published *Flora Malesiana* (1950) to mark the international achievement of establishing the region of Malesia. The very basis of the work had been "laid by representatives of many nations," with its scope "following the limits imposed by Nature rather than those made by man." In other words, the geographical area or phytogeographic space covered by Malesia ("Malaysia," as van Steenis spelled it) included "more territory than is ruled by a single Government."[9] For van Steenis, completing such a comprehensive work required cooperation among those stationed in colonies and metropoles and the assembly of decades of data collected by foreign and local plant specialists. In the volume, he describes Malesia as a natural geographic unit with a history that began in the twentieth century, although he credits the work of collectors and botanists in the previous century.[10] He especially celebrates Merrill, who in his view had "started without a predecessor from absolute scratch, without personnel, without a book or collection, in an almost unexplored very rich archipelago covered largely by primary forest."[11]

One scholar of biogeography has suggested that colonialism, "a strong driving force in regionalization," has influenced the growth of regionalization's intellectual history by helping historians identify those who "collected specimens and measurements."[12] But suggesting that the "aims and goals" of biogeography were not impacted by colonialism fails to account for the very infrastructure, deployments, intellectual commitments, and conditions by which researchers—foreign and local alike—could engage in work that would contribute to the knowledge of floristic regions. To deny the influence of colonization on Vidal's or Merrill's efforts warps the political and social features of science and how a colonial setting shaped them. This denial obscures how knowledge of flora abetted colonial ambition. Most counterfactual of all is the notion that only collectors and measurers could emerge from a colonized place to sustain the empirically comprehensive work of regionalization. Such an assertion, as I further argue below, coincides with what Jorge Cañizares-Esguerra has called an "epistemological colonialism and the cult of [Anglo-European] saints" founded on the supposition that "[t]he global South has tarantulas, the North ideas."[13]

By the early twentieth century, botanists had already begun to delineate the geography of that which would become Southeast Asia before the social scientific emergence of the region. A similar propensity cropped up in the discipline of ichthyology and among local intellectuals, as Anthony Medrano has convincingly shown.[14] In the intellectual history of Southeast Asian studies, the term and region "Southeast Asia" gained popularity at the time of *Flora Malesiana*'s first publication. This is not to say that "Southeast Asia" (spelled, hyphenated, and capitalized variously) was not used decades or even the century prior. But it was "warfare not scholarship" that gave the region currency: "Making war meant making maps" goes the well-known Donald K. Emmerson quote.[15] World War II not only occasioned the popularization of the term but also standardized the geographical boundaries of the region. Intellectual production followed and began to assert the region's historical and cultural coherence. Scholars in the mid-twentieth century historically and culturally mapped the region in various ways, at times including or excluding the Philippines, for instance. Thinking regionally, however, was not limited to the domain of the social sciences, whether at the turn of the century or into the mid-twentieth. Regional thinking clearly preoccupied colonial botanists, as well. They and their collaborators mobilized Philippine flora to continue to realize the aspiration of a global Linnaean systematization on newer geographical terms.

In their shared objectives, Spanish and US colonial botanists were never innocent of the environmental extraction and inequities within collaborative relationships that characterized the science in the Philippines. It would seem that in their pursuit of intellectual advancement, foreign botanists appeared to stand outside of—or keep mum on—the colonial violence and wars that surrounded them. They operated in a global atmosphere of high ideals amid trenchant colonial expansion. The IBC's gradual evolution occurred alongside the enactment of internationally oriented agreements that sought to dissuade, however impotently, violent imperial contests. These agreements, such as the Berlin Conference (1884–1885) to regulate trade and colonization in Africa, the Brussels Conference Act of 1890 to end land and sea slave trade in parts of Africa and the Ottoman Empire, the Paris Peace Conference (1919–1920) at the end of World War I, and the 1920 establishment of the League of Nations comprised the commercial, political, and military landscape of new imperial formations in the early twentieth century.

Recuperating Actors, Recuperating Ways of Knowing

A more symmetrical history of the US and Spanish regimes allows for a clear-sighted recognition of continuities fundamental to a shared scientific enterprise. These continuities were human and vegetal. My investigation of the broader intellectual, institutional, cultural, and political contexts surrounding the science of botany has also permitted more historical actors to take the stage, so to speak. Their labor and intellectual production force historians to confront several of the US critiques of nineteenth-century science in the Philippines that were not only targeted at the Spanish but also at Filipino personnel.

Paramount in my investigation has been García. His interest in the colony's extensive rice varieties, seniority in the colonial service, and artistic capacity made him essential to the success of Spanish (and eventually US) botany and to recording vernaculars in the archipelago. His racial and class position, as well as his linguistic skill, enabled him to log the "fluctuating form." His catalogs for the JBM point to plant knowledge not fully communicated through *Oryza sativa* alone. For his *Sinopsis*, his position as an artist in elite Manila led to the creation of a visual archive of plants that was both aesthetically bold yet still instrumental to the furtherance of colonial botany. His acumen was born of the capital, which equipped him with techniques that he brought to illustrating plant life. García's professional skills remained apposite through the Philippine–American War and the start of the US colonial period, when he provided institutional know-how for two disparate camps.

Merrill appreciated García, whom he met early into his stay in Manila. Beginning in 1900, with the future of the First Philippine Republic unclear, García collaborated with the US. He entered US government service in May 1900, ahead of General Arthur MacArthur Jr.'s (1845–1912) June 1900 amnesty proclamation for all who renounced the republic and pledged allegiance to the United States.[16] He supplied institutional memory to the newly arrived corps of US plant researchers, and his recollections served in lieu of extensive documentation detailing Spanish operations, so much of which was lost in a fire at the JBM in 1897.[17] Not unlike some the work he performed on behalf of the Spanish for over three decades, his efforts assisted in institutional transition. In the first years of their professional relationship, Merrill wrote of García, "Although, much of his work might be criticised, still, considering the training—or rather, lack of training—he has had it is better to suspend judgment."[18]

George P. Ahern, one of Maximo Ramos's earliest US collaborators, also looked to García for his familiarity with Philippine forests and the IGM. Ahern loudly derided Spanish scientific work, though he needed to rely on the IGM's infrastructure to erect similar institutions on behalf of the United States.[19] He was skeptical of García's authority. In 1901, Ahern published notes on Philippine woods for which he compiled direct quotations from previous Spanish-era publications and comments and illustrations from and by García. At the end of the compilation, Ahern provided biographical sketches for the authorities cited, like the Vidal brothers, Manuel Blanco, and Henry Brown, an English timber magnate.[20] García had no biographical sketch, even though his knowledge and floral illustrations permeate the publication. To Ahern, García may have been competent but he was not an adequate botanist: he allegedly made guesses, possibly with respect to plant identification.[21] Separately, another US forester referred to García as "only an old Spanish botanist" with no facility in the English language.[22] In 1903, Merrill recognized that García had established an herbarium at the Ateneo Municipal in 1894 that contained roughly 1,000 specimens. But without labels indicating the growing localities or dates of collection, García's herbarium was "of no historical and little scientific value."[23]

In light of such criticisms, as Nathaniel Roberts interprets, García "bore the dual taints of 'Filipino capacity' and experience in the [IGM] and therefore did not fit Americans' idealized model" of a competent government official.[24] Careful study of Spanish botany operations nonetheless reveals García's fundamental role in the production of knowledge about the Philippine environment. Even with the near complete obliteration of JBM archives, surviving ledgers kept in Madrid point to island-born men who extended the institutional capacity of Spanish colonial science. More symmetrical scrutiny identifies more precisely where locally born individuals contributed to the Spanish outfit, and more damningly, how US critiques were leveled against peninsular and insular people alike.

What continues to be heralded and remembered as the US administration's "golden age of botany" must reckon with the historical emergence of the title and the local actors who are forgotten in the celebration of discretely US achievements. The trend of US "imperial renewal," which could reframe the US as "reformers who would sweep away an antiquated, corrupt, Iberian colonialism and replace it with a modern, vigorous, Anglo-Saxon administration" was present in botany, too.[25] Yet, as others have written, the "American imperialists were inheriting a series of late Spanish imperial states that were in the midst of their own liberal renovation," as they were "markedly more effective and expansive than the government in Spain itself."[26] The staying quality of US colonial-era critiques hides the social nuances of intellectual production of the time, the originality captured within publications, and the historical contingencies that determined authorship, employment, and access to vernaculars.

This book has also taken seriously botanical work within the milieu of ilustrado production. Excellent scholarship has examined the late nineteenth century disciplinary vehicles through which ilustrados asserted their knowledge of the colony. Botany had a place within this native intellectual fervor. García put forth knowledge-claims to the Philippine environment as his contemporaries did in what may be considered humanistic and social scientific fields. The Filipino Enlightenment, of which García's oeuvre should be considered a part, spilled into the early decades of the twentieth century when the US regime most needed local support. García was not the only educated, local Philippine actor to work through the imperial transition. Trinidad Pardo de Tavera, the pharmacist León María Guerrero (1853–1935), Escuela de Agricultura graduates, and IGM foresters, to name a few, carried their knowledge of flora into the daybreak of the new colonial regime. People like Ramos and

Oleng lived through the transition, and their insight, though of use to US researchers, evidences fuller capacities of plant knowing that predated the US regime. Future targeted studies of personnel should critically examine other continuities and potential discontinuities to reveal even more insight into plants than this book has covered.

Recuperating local actors in colonial history is not a new endeavor. I follow dozens of remarkable historians who have done similar work. In my Philippines case study, recuperation has facilitated the very empirical and conceptual foundations of the sovereign vernacular. Historical actors demonstrated botany's continual making and unmaking: the edges of the science frayed by superstition and inspirited fabric; and the edges hardened by local labor and a deft knowing of the environment. They showed that in spite of botany's internationalizing trajectory, the science and its practitioners had to contend with knowledge of plants that simply would not fit their own parameters of interest and epistemological limits.

Following particular plant species has also enabled a better understanding of how species changed, developed, and transformed within novel grids of meaning. Dynamic in its own right, the environment offered the material intra-active capacity across the regimes. At the same time, plants did not just summarily disappear during the colonial transition. What was perceived to be the same material species across 1898 allowed for sustained study of plants and yet also showed the variable registers a plant could assume depending on the colonial order. This is most especially the case with the sampaga and the sampaguita, a plant that Sophie Chao, applying the work of Susan Leigh Star and James R. Griesemer, would describe as "accru[ing] meaning as it travel[ed] across spatial and conceptual realms, drawing together disparate individuals, communities, and institutions."[27] The species' symbolism shifted depending on the local intellectual and political climate, only to be transmuted, under US auspices, into the national emblem it is today. Transformations appear in woven objects and tools from southern Mindanao, signifying cultural, social, and economic shifts with the rise of abacá plantations. The banana fiber that had been a craft material and a valuable export good in the preceding century became part of the backbone of the US colonial economy in the Philippines. Its commercial weight attracted pioneering foreigners, domestic and international, to seize upon a presumed frontier. In the 1950s, the abacá economy crashed due to the introduction of synthetic fibers to the global market and to pest blights that hit plantations.[28] As this book has emphasized, plant species deserve even more heightened analysis to reconsider Philippine history through a fresh lens. More remains to be done.

Sovereign Vernaculars

How should historians investigate and problematize the ways we come to know plants? Sovereign vernaculars, as I have laid out, should remind scholars that the ways by which we develop knowledge of plant life are born of place, may be impermanent, and can live in even more materials yet to be examined in the history of science.

Place matters. This book has shown that the location where plant knowledge emerges has important bearing on that knowledge. In concordance with Thomas F. Gieryn's argument, truth-claims and place are inextricable.[29] For the Philippines, this argument could be seen among rice varieties. Varietal names traveled to and from Manila markets and to García's seed bank at the JBM, revealing historical actors' more complete experience of rice—from its taste and form to its use and cultivation period. In García's seed catalog, dinulung—a Visayan word for both a rice variety and a small, brownish anchovy—suggests an equating of semi-aquatic and marine life forms. Tiaong, also from the list, was a town name in Tayabas, likely indicating the variety's provenance. Also from the Visayas and the Cebuano language, was bihod, a word for fish roe that could perhaps mimic the consistency of the variety made roundish and shiny after boiling.[30] For an archipelago rich with dozens of languages and dialects, any number of place-specific factors could have informed these varieties' names. What they show are exciting social worlds that surround and are made by a single grain of rice.[31]

The place of plant knowledge can also reveal the power structures at play when a nationalizing force, for instance, comes to proclaim a single flower as a national symbol. Though the 1930s cemented the Manila sampaguita's countrywide status, its cultural importance was nevertheless forged in the nineteenth century. In the crucible of politics and intellectual production of the time, the flower had profound gravitas that would allow it to be taken up and mapped to the rest of the colony, even when other names associated with the plant were uttered elsewhere. As Vicente L. Rafael has written, "[N]ot only is nationalism inherently conflictual, caught between dynastic/colonial modes of apprehension on the one hand and the possibilities of an egalitarian, postcolonial existence on the other; but that the means for imagining nationhood may at times be at odds with the very nature of the images that are reproduced."[32] For Rafael, popular and heavily reproduced images of José Rizal promote a type of nationalism that "flattens" the complexity of Rizal's writings, outlook, and persona. "Like 'tenacious phantoms,'" Rafael writes, "his images have clung to the national consciousness."[33] I gesture the same for the sampaguita, a flower with historically immense meaning whose deployments in the late nineteenth century were part of much more politically revolutionary aspiration and fervor than what the civic nationalism of the US colonial era decreed. A kind of sanitizing took place, stripping the flower of the very contexts in which its political invocations were loudly broadcasted.

More recently, Philippine lawmakers have tried to raise the waling-waling, an orchid endemic to the island of Mindanao, to national status. In 2013, then President Benigno Aquino III vetoed the bill that sought to elevate the waling-waling's repute. His veto claimed that the flower's declaration would have "the pernicious effect of displacing the hallowed status of the Sampaguita, a cherished national icon, as the primary symbol of Philippine culture and artistry."[34] In 2020, Philippine lawmakers again tried to push for the waling-waling. When asked his view of the debate, the then-director of the National Museum of the Philippines agreed that a second flower would be suitable but that it could not replace the sampaguita's "ubiquity and cultural value."[35] The "cultural icon" could not be swapped in spite of its limited growing patterns, lack of endemicity to the archipelago, and ties to the US state-making project. These sorts of debates are indeed born of place, recognizable because of the domestic regional politics of which they are a part. The waling-waling undoubtedly raises questions over the role of Mindanao in the national imaginary and the sampaguita's nativeness. These concerns point to the sovereign vernacular: even if the waling-waling is understood by botanists to endemically come from the Philippines, unlike the Arabian sampaguita, such a justification does not outweigh the flower's cultural import and place in the national body politic.

For weavers in Santa Cruz, place could yield just the right color, just the right tactility to a fiber, and just the right opportunity to include materials—store-bought or plant-derived—into finished fabrics. For particular communities, places sprouted demons, spirits, and dwarves, some "legion," some exclusive to the plants in which they resided. These considerations juxtapose the ways in which plant life of the Philippines could be generalizable to commercial markets, to herbaria collections, and to botanists' regionalizing tendency. An abstraction of plants from place can be observed across the colonial botany material, and such abstraction counters the contextually embedded nature of certain plant knowledge arising from the Philippines. Deep considerations of place—ones that can surely be built upon and go *even* deeper—counter the mega-narratives that colonial botanists viewed themselves producing and at the same time, demonstrate the local circumstances within which botanists worked and from which they derived such mega-narratives.

Moments matter. Plant knowledge changes, and my preoccupation with time in this book has focused on the relationship between fixity and impermanence. I hold to the belief that sovereign vernaculars should be understood in *moments* to push against the idea that fixed knowledge is the most relevant and worthy of documentation. This book has shown that notions of plants could change, some seemingly so inconsequential that botanists either considered them unreliable or did not record them themselves. In this sense, I have urged readers to consider moments when the project of a fixed idea of plants—whether at the hands of botanists or local Philippine actors—could be upbraided at any time.

I affirm that some plant knowledge may never have been captured, whether in writing or epistemically, by the colonial botanists I have covered. This affirmation holds for the knowledge systems relegated to the realm of superstition, for instance. Although some may argue that such knowledge may have been beyond botanists' intellectual objectives, the very narrowness of such objectives has led me to posit the told and untold manners by which peoples knew plants. That botanists would come to cast such knowledge to other intellectual disciplines, such as anthropology, continues a lineage begun in Europe to erect such precincts of thought. This book has insisted that studying the finished records of colonial Philippine botany demands examining the discipline-making that came along with it, thus provincializing the science.[36]

Assumptions of the power of fixed knowledge, what's more, should be acknowledged within taxonomic debates today that seek to weed out the colonial vestiges found in Latin names and to replace them with indigenous monikers.[37] To be clear, an effort to rename patronymics is not "politically motivated censorship and totalitarian cleansing of scientific history," as one scholar has alleged.[38] As I have shown, botanists may continue to practice synonymy, a (babbling) historical record in and of itself.[39] Widespread public precedence today has insisted scientists reckon with the checkered histories of their disciplines. Patronymics that laud the conceits of colonial figures are haunting. Still, the very existence and global use of Latin binomials are inescapably part of colonial history.

To revise binomials with Latinized indigenous or local nomenclature, while an admirable corrective, reenacts a universalizing of plant names to which a global community of users is expected to adhere. The logic of fixity behind indigenous or local names that may supplant Latin nomenclature points to the issue of the sovereign vernacular that is both sovereign to the objectives of botany and yet part and parcel of its dominance. Communities of speakers may continue to powerfully recite plant names spoken centuries ago. A sovereignty emerges in these very names, which have not been expunged by Latin's touch.

Yet before fixing such names within the project of Latinate botany, proponents of nomenclatural revision should acknowledge that such a move universalizes knowledge once again, leaving communities to contend with perceptions and enactments of a hierarchized vernacular.

Finally, sources matter. Sovereign vernaculars are not only a concept but also a possible method for thinking through the history of science. A variety of source materials, those seemingly external to the ecosystem of a conventional history of science, can help tell a different history, one among several that challenges the narrative preeminence of imperial or colonial botany and their cast of usual suspects.

Respect for varied source material does not lead to a satisfactory systematic result, at least not completely fulfilled in the Linnaean dream. That historical actors could produce columns on gardening, that plant names could change, that artistic practices could break from convention, or that poetry could be uniquely Ilocano or not, shows the disobedient, unexpected, and yet perfectly sensible terms by which humans come to engage with plant life. These terms overpower the sense of botany as the preeminent manner of knowing plants. Indeed, this opinion was held in the late nineteenth century by a columnist for *Revista de Filipinas*, who in July 1876 published on the mabolo, caturay, and sampaloc. The columnist aimed to write on Manila's trees but included Latin nomenclature and admitted that its inclusion was "pedantic" were it not for *Revista*'s international readership.[40] Historical narratives emerge in the material, ones that show a more complex dynamic that sustained, if even momentarily, the number of ways people knew plant life despite (and sometimes because of) the science of botany.

I continue to invite readers to recognize the presence of vernaculars in varied sources, their existence during a significant period in the history of imperial science, and their impact. Other source material exceeds the scope of this book and can make for more incisive history. Horticulture, culinary traditions and cuisine, healing craft, religious practice, and in-depth lexical study can shed more light on ways of knowing plants. Other languages and islands promise to reveal even more than what my Tagalog-heavy study has uncovered. In the Philippine context, scholars have examined botanical research undertaken by religious orders, documentation on medicinal plants, and agricultural methods pursued by ethnolinguistic communities.[41] Some are employing exciting interdisciplinary tools to study land use and plant exchange well before and during the colonial encounter.[42]

With my material at hand, I have inherited Rafael's work on the vernacular and have read for "subaltern agency brushing against the grain" of colonial aspirations.[43] His approach motivates me to return to the lantern slide of the field assistant on the mature balete, the image that opens this book, to read the knowing within it. I have hesitated to cast aside the photograph as only a colonial remnant, one that might too readily imply the assistant as passive or silent.[44] Likewise, I have refrained from interpreting agentic moves that may be too imposing given the primary material I have discussed. I have also refused the inclusion of other colonial images, whose explicitly violent, racially categorizing, or demeaning imagery need not be reproduced to define a colonial place long permutated through such representations. Instead, I have selected the lantern slide to demonstrate how historical circumstances conditioned its production and how it may challenge some in/visibility metaphors that can riddle studies of colonial science. The field assistant reclining on the balete once had more knowing about plant life than I can adequately surface. His knowing, and those of others, may live beyond historical source material; may be permanently obliterated by colonial violence. I honor the knowing that once was and that which I may never recover.

Acknowledgments

This book is a result of my father, my mother, and our forebears. If it just so happens that children choose their parents, I chose mine because I knew they could show me a thing or two about plant life and quiet tenacity, written creativity and pluck. Hermes Garces Gutierrez is my kindred spirit and my interlocutor for this project and the ones to come. Estrella Balbieran Cruz is a force of nature, who reminds me to take no prisoners, most especially when you're under five feet tall.

While my parents gave me my start, my academic mentors guided me along the way. Peter Zinoman, Penny Edwards, the late Jeffrey Hadler, Catherine Ceniza Choy, Lisandro Claudio, and Massimo Mazzotti have invested considerably in me. They've molded me into a better scholar, a better writer, and really, a better human. Peter became my primary adviser during a time of searing loss and continues to be my best critic. I thank Penny and Jeff for noticing some spark in me long ago in their undergraduate classrooms and for opening a path in the academy when I hadn't known one existed.

Jackson Perry, astute reader and scholarly companion, got me through the ugliest part of the drafting process. My cohort at the Oak Spring Garden Foundation, particularly Manjula Martin and Joseph Zordan, softened the moments around my ascetic work hours. At the University of California, Santa Cruz, Caitlin Kelliiaa and Yasmeen Daifallah were the right kind of writing partners: equal helpings humble and sharp. Patricio Abinales, R. Baker, David Biggs, Ania Mah Gricuk, Erik Harms, Gail Hershatter, Thiti Jamkajornkeiat, Orven Mallari, James Marks, Tri Phuong, Noelle Sepina, Adriane Kelani Stoia, Kristine Swarts-Zanin, and Joseph Zordan generously read my in-progress chapters. Jorge Cañizares-Esguerra, Vicente Rafael, Banu Subramaniam, Megan C. Thomas, and Lisandro Claudio tackled the full manuscript with me to get it into shape.

Ken Wissoker, Kate Mullen, and Michael Trudeau at Duke made the publishing process windowpane clear. I thank Ken for eyeing something promising in this work. I owe heaps of gratitude to the two anonymous reviewers who made the book sing. Truly, any thin analyses, omissions, and remaining errors are my own.

I had the privilege of working with researchers who assisted me at various stages of this project. From De La Salle University, Manila, Johanna Gatdula, Jaime Consulta, Keziah Aurelio, and Nina Ty digitized and located some of the most significant material upon which this book relies. Joey Baquiran, Cristina Guillén Arnáiz, and Phi Yen Nguyen were vigilant reviewers of my Tagalog, Ilocano, Spanish, and Vietnamese translations. In Santa Cruz, Owen Cooksy helped me format the final manuscript and practiced careful reading of it. John Wyatt Greenlee of Surprised Eel Mapping designed the map that graces this book's introduction. Academic Mechanic, Jordan Beltran, prepared the thorough index. Johaina Katinka Crisostomo, Kashi Gomez, Lawrence M. Kelly, Šebestián Kroupa, Sohini Pillai, Michael Purugganan, Cherubim Quizon, Emily Sessions, and Lincoln Taiz were subject specialists who indulged my questions. I must credit Anthony Medrano and Luthfi Adam for helping me phrase the succinct definition of sovereign vernaculars. What one can learn over tea and a couple of pastries.

Since I was a wee Southeast Asian studies major at Berkeley, Virginia Shih has provided prompt research support when I've been in need. Paul Michael L. Atienza, Christianne F. Collantes, Michael Hawkins, Karen Buenavista Hanna, Ryan Nelson, Joseph Scalice, and Noah Theriault stay reliable friends and fonts of expertise. In Santa Cruz, balancing the community-engaged research initiative Watsonville Is in the Heart and this book wasn't easy, but the Watsonville team—and I mean *team* in every sense—made it doable: Steve McKay, Meleia Simon-Reynolds, Christina Ayson Plank, Dioscoro "Roy" Recio Jr., and the effective force behind the Humanities Institute, Irena Polić, Saskia Nauenberg Dunkell, and Jessica Guild. Christina deserves a special shout-out: during my hardest months spent revising, she worked with discipline on our successful grant application to the National Endowment for the Humanities. Our undergraduate researchers inject energy into my days, and I pray only the best for them, especially Ian Hunt Doyle and Maia Jardenil Mislang.

My work was made realizable by lengthy research stints that afforded me time and space to think and to discover. In the Philippines, I thank Esperanza Maribel Agoo, brilliant plant researcher and my faculty mentor at De La Salle University, Manila. Because of her I met the team behind the *Philippine Journal of Systematic Biology* and the Association of Systematic Biologists of the Philippines, to which all proceeds from this book's sale will go. I am indebted to Domingo Madulid, historical botanist par excellence. Our conversations have been the most thought-provoking and have forced me to consider the botanizing clerics with serious care. The Miguel de Benavides Library at the University of Santo Tomas (UST), the American Historical Collection at Ateneo de Manila University, and the Mario Feir Collection were key Philippine repositories for this book, as well as the image banks of the Lopez Museum and the Filipinas Heritage Library at the Ayala Museum. At UST, Diana Padilla and Ginalyn Santiago showed me hospitality in the archives. In the University of Philippines system, I value Sanley Abila, Kerby Alvarez, Ros Costelo, Ruel Pagunsan, and Analyn Salvador-Amores as dependable collaborators and friends. In Madrid, I had the matchless opportunity to work with Florentino Rodao, who made the Universidad Complutense de Madrid my institutional headquarters and the capital city, a welcoming place. The archivists at the Archivo Histórico Nacional made research easy if not pleasurable.

In the United States, I thank Daniel Lewis at the Huntington Library, who unlocked storytelling in environmental history for me. Vanessa Sellers, Stephen Sinon, Samantha D'Acunto, and Shae Werth made the New York Botanical Garden a cozy workspace. The real work to produce this manuscript happened in their presence even when I made the ill-advised decision to quit caffeine partway through the process. Through them, I met Kate Armstrong, Douglas Daly, Peter Fritsch, Lawrence M. Kelly, and Matthew Pace, plant researchers who answered my questions despite the demands on their time. While in New York, I met Cherubim Quizon, who practiced tremendous scholarly charity with me. Because of her, I followed a trail of pandesal crumbs to the American Museum of Natural History and the Smithsonian Institution to find beloved textile holdings. I appreciate Mary Lou Murillo for her pleasant help as I examined such collections, and Patricia O. Afable and Joshua A. Bell as I tried to write coherently about them. Michael Purugganan and Elaine Gan provided me with community in the Big Apple, and I remain awestruck by their inventive research. Because of Juno Salazar Parreñas I stopped through Ithaca, and while at Cornell had the chance to meet Jeffrey Peterson, whose insight helped me comprehend the ontological turn. In Michigan, Diana Bachman located important holdings at the Bentley Historical Library. Mike G. Price, like a real tito, fully welcomed me to Ann Arbor and his stories of his time with my dad under Marcos's martial law were comical albeit unsettling.

My graduate research in the US, the Philippines, and Spain, upon which much of this book is built, was made possible through grants from the Center for Race and Gender at Berkeley; the International Doctoral Research Fellowship through the Social Science Research Council funded in part by the Mellon

Foundation; the Doctoral Dissertation Research Abroad fellowship of the Fulbright-Hays program of the US Department of Education; the Dibner Fellowship in the History of Science and Technology at the Huntington Library; and the Bordin-Gillette Research Grant of the Bentley Historical Library. Foreign Language and Area Studies fellowships from the Center for Southeast Asia Studies and the Institute for European Studies, as well as a travel grant through the Council for International Educational Exchange, enabled my language studies in Madison at the Southeast Asian Studies Summer Institute and in Seville in partnership with the Universidad de Sevilla.

As faculty at UC Santa Cruz, I have been funded by the Humanities Institute's Research and Publication Grant, the Committee on Research, and the Society of Hellman Fellows. Generous support for the publication of this book was also provided by the First Book Subvention Program of the Association for Asian Studies. I couldn't have churned out a semblance of a manuscript if it weren't for the time the university gave me so I could complete two significant residencies: a nine-month postdoctoral fellowship funded by the Mellon Foundation at the Humanities Institute of the New York Botanical Garden and an interdisciplinary stay at the Oak Spring Garden Foundation. I spent many days thinking and writing (and writing and thinking) at these places. How crucial they were.

The content of this project was enhanced by the number of audiences who offered critical feedback. Over recent years, these audiences have included the Department of Botany at the University of California, Riverside; Cultural Studies Colloquium at the University of California, Santa Cruz; Southeast Asia Program at the University of Washington, Seattle; Gatty Lecture Series at Cornell University; Center for Southeast Asian Studies at the University of California, Los Angeles; Center for Asian Pacific American Studies at Pitzer College; Southeast Asian Lecture Series at Harvard University's Asia Center; Sulo: Global Philippine Studies Initiative and Experimental Humanities and Social Engagement at New York University; Center for Southeast Asian Studies, University of Wisconsin, Madison; Archaeological Studies Program and the Science and Society Program at the University of the Philippines, Diliman; Macmillan Center Council on Southeast Asia at Yale University; Departments of History and Botany at Wells College; Botanistas and Club Filipina at Wellesley College; and Lembaga Ilmu Pengetahuan Indonesia (now Badan Riset dan Inovasi Nasional) and Ritsumeikan University. I thank the attentive ears I've come across in these spaces.

The Department of History and the Humanities Division at UC Santa Cruz have given me ample cognitive latitude to complete this project. Matt O'Hara,

Alice Yang, Jasmine Alinder, and Jody Greene have demonstrated much leadership and understanding for junior scholars like me. For their regular guidance, I appreciate my campus mentor, Felicity Amaya Schaeffer, my mentors in History, Jennifer Derr and Kate Jones, as well as members of the History Department's Asia Caucus, Noriko Aso, Shelly Chan, Alan Christy, Minghui Hu, and Juned Shaikh. The Center for Southeast Asian Coastal Interactions (SEACoast), namely Megan Thomas and Anna Tsing, opened the door to a beautiful early career on the Central Coast. Through SEACoast, I've gotten to think with the plant-minded virtuosity of David Biggs and Sophie Chao. I thrive in Santa Cruz because of colleagues like Shelly Chan, Amy Mihyang Ginther, Alma Heckman, Caitlin Kelliiaa, Kailani Polzak, Amanda Smith, Elaine Sullivan, and Zac Zimmer.

I must have done something right in a previous life to have been given star siblings and longtime friends. Without them, I would be hollow. Kimler Gutierrez, Kamille Mosqueda, Kissette Mosqueda-Kelly, and Scott Kelly bless my days. Kamille has walked closest beside me on this protracted career path and is usually the first person I call when I experience a win or a loss. Victoria Abcede, George Borgona-Chacon, Frances Borgona-Chacon, Alison Chopel, David Dao, Jack DeJesus, Fayzan Gowani, Benjamin Gunning, Irene Headen, Soobin Kim, Yuriy Mikhalevskiy, Antonia "Toni" Martinez, Marcus Poon, Allan Sahagun, Jannice Saniano, Sunshine Velasco, and Megan Zapanta make time on this planet something I can't get enough of. Carol Charlton, Janine Gordon, Joey Gordon, Eileen Murray, and the Pates, Becky, David, and George, make the winters in Florida restful. I appreciate Carol, who, among all my personal circles, asked me the most about my publication tidbits.

My husband, Ryan Pate, deserves more than these typeset words can impart. He is smart chord changes and a swatch of bright watercolor. He is contentment found spooning on a living room couch. He is my peace. Ryan had to deal with this book in all its bedraggled and confused forms; weather my complaints, my anguish, and my early mornings spent rushing to a keyboard to eke out a sentence. How he found the patience through all of that, I don't know. But such is love.

Notes

INTRODUCTION. SOVEREIGN VERNACULARS

An earlier version of portions of this introduction appeared in *History Compass* as "From Objects of Study to Worldmaking Beings: The History of Botany at the Corner of the Plant Turn." I thank the three reviewers with *History Compass* for recognizing the potential in plant worldmaking.

1. Myers, Mittermeier, Mittermeier, de Fonseca, and Kent, "Biodiversity Hotspots for Conservation Priorities," 853–58; "Philippine Biodiversity," *Expeditions at the Field Museum*. I thank Peter Fritsch for these references.
2. Pelser, Barcelona, and Nickrent, Co's Digital Flora of the Philippines website.
3. Orillaneda, "Maritime Trade in the Philippines during the 15th Century CE," 83–100.
4. Mojares, *Brains of the Nation*, 381–466.
5. Joaquin, *A Question of Heroes*, 5.
6. Claudio, *Jose Rizal*, vii.
7. Schiebinger, *Plants and Empire*.
8. For a comparative examination of Spanish and US racial regimes and how these influenced Cuban antiracist nationalism of the turn of the nineteenth and twentieth century, see Ferrer, *Insurgent Cuba*. For the Philippines, an effective overview of US educational and bureaucratic policies after US takeover of the colony can be found in Abinales and Amoroso in *State and Society in the Philippines*, 113–25.
9. Thomas, *Orientalists, Propagandists, and Ilustrados*, 40–44.
10. See, for example, the chapter "The Undead," in Rafael, *White Love*, 77–102; Vergara, *Displaying Filipinos*; Rice, *Dean Worcester's Fantasy Islands*; and Balce, *Body Parts of Empire*.
11. Harley H. Bartlett, "Nationalism, Imperialism, and Spheres of Influence in Natural Science," manuscript, pp. 4–43, folder 3, Concerning Botanical Subjects, box 7, correspondence, Harley H. Bartlett Papers (hereafter BHL).
12. W. Anderson, "Science in the Philippines," 288. The historiographical tendency can be found in Agoncillo, *Prelude to 1896*, 2. See also Agoncillo, *Introduction to Filipino History*; Zaide, *History of the Filipino People*, 220–21.
13. In addition to "Prescott's paradigm," coined by Richard L. Kagan as the trope of "Spain as America's antithesis," other scholars have addressed the impact of "la leyenda

negra" ("the black legend") on Spanish imperial studies and efforts to revise histories of the European science. See, among others, Kagan, "Prescott's Paradigm," 423–46; Eamon, "'Nuestros males,'" 13–30.

14. The work of Bankoff has been especially vital to this historiographical correction. Among his writings, some of the most instructive include Bankoff, "A Month in the Life of José Salud," 8–47; "The Science of Nature and the Nature of Science," 78–108.

15. For coverage of this historiographical trend and its implication for colonial science studies on the Philippines, see Bankoff, "The Science of Nature and the Nature of Science," 78–79.

16. Restrepo and Rojas, *Inflexión decolonial*, 140–41, as partially translated in De Lima and Schuster, "Decolonizing Global History?," 444. See also W. Anderson, "Science in the Philippines," 289.

17. Englund, "Ethnography after Globalism," 267. See also Casey, "How to Get from Space to Pace in a Fairly Short Stretch of Time," 39.

18. Santos, Nunes, and Meneses, "Opening Up the Canon of Knowledge and Recognition of Difference," li.

19. Banu Subramaniam, "Science and Postcolonialism."

20. Barbour, "Embodied Ways of Knowing." See also Anderson's deft treatment of "feminine 'ways of knowing'" vis-à-vis feminist epistemology in "Feminist Epistemology," 61–66. I use "ways of knowing" as a feminist methodological stance rather than an assertion of essentialized difference between male and female knowers and knowledge-producers.

21. Important readings on this include Schaeffer, *Unsettled Borders*; Taylor, Moggridge and Poelina, "Australian Indigenous Water Policy"; Whyte et al., "Weaving Indigenous Science." See also Salvador-Amores, *Anthropological, Mathematical Symmetry and Technical Characterisation of Cordillera Textiles Project*, which firmly considers Cordilleran weaving practices as scientifically legible.

22. Watson-Verran and Turnbull, "Science and Other Indigenous Knowledge Systems," 116; Daston and Galison, *Objectivity*.

23. Schwartzberg, "Cosmography in Southeast Asia," 714–15.

24. *Mahābhārata Book Twelve*, 105–7.

25. Green, "Angkor Vogue," 448.

26. Barnes, "An Introduction to Buddhist Archaeology," 165; Ryan, "Banyan," 23.

27. Xhauflair et al., "The Invisible Plant Technology of Prehistoric Southeast Asia."

28. Ho and Lisowski, *A Brief History of Chinese Medicine, Second Edition*, 31–33.

29. Greek herbals took hold as an important genre for recording plants, and many histories of European botany laud Pedanius Dioscorides (ca. 40–ca. 90) and his five-volume *De materia medica* as precursory to botanical science.

30. Esposito, *The Oxford History of Islam*, 212. See also Dallal, *Islam, Science, and the Challenge of History*, 43.

31. Margaret, email communication with author, November 18, 2021; Ho and Lisowski, *A Brief History of Chinese Medicine*, 32; George, "Direct Sea Trade between Early Islamic Iraq and Tang China," 609.

32. Monnais-Rousselot, Thompson, and Wahlberg, introduction to *Southern Medicine for Southern People*, 1.

33. Reyes, "Glimpsing Southeast Asian Naturalia in Global Trade, c. 300 BCE–1600 CE," 108.

34. Reyes, "Glimpsing Southeast Asian Naturalia," 105. Reyes cites and usefully engages Wheatley, "Geographical Notes on Some Commodities Involved in Sung Maritime Trade."
35. Ray, *Climate Change and the Art of Devotion*, 101–2.
36. Ogilvie, *The Science of Describing*, 231–58.
37. Butzer, "From Columbus to Acosta," 545. See also Kelley, *Clandestine Marriage: Botany and Romantic Culture*, 5. Kelley cites Mitchell, "Cryptogamia," 631–51.
38. Sigrist, "On Some Social Characteristics of the Eighteenth-Century Botanists," 205–11.
39. Bonneuil, "The Manufacture of Species." Bonneuil refers to Lesch, "Systematics and the Geometrical Spirit," 73–111.
40. Cooper, *Inventing the Indigenous*, 166–72.
41. Ogilvie, *The Science of Describing*, 218.
42. For Linnaeus, plants that did not display clear sexual organs, like fungi or algae, became "cryptogamia," or those who married in secret. See Moore, "Linnaeus and the Sex Lives of Plants," 132. See also Müller-Wille, "Linnaeus and the Love Lives of Plants," for Müller-Wille's excellent reading of the cameralist politics behind Linnaeus's anthropomorphized approach to plants. On cultural bias and the centuries-long debate over plant sexuality, see Taiz and Taiz, *Flora Unveiled*. For an updated commentary on the legacy of Linnaeus's system, see Subramaniam and Bartlett, "Re-imagining Reproduction."
43. Sigrist, "On Some Social Characteristics," 211–22.
44. Drayton, *Nature's Government*, 55–67.
45. Schiebinger traces the biopiracy of Nicolas-Joseph Thiéry de Menonville (1739–1780), a French botanist who sought to acquire and propagate the cochineal beetle from Spanish-colonized New Spain for its valuable crimson dye. Thiéry de Menonville, according to Schiebinger, succeeded others such as Pierre Belon (1517–1564), Pierre Poivre (1719–1786), and Jean-Baptise Leblond (1747–1815), who also poached prized plants from rivals' colonial terrain. See Schiebinger, *Plants and Empire*, 35–46.
46. Mackay, "Agents of Empire," 38–48.
47. Bonneuil, "The Manufacture of Species," 190–91. See also Secord, "Corresponding Interests."
48. Bonneuil, "The Manufacture of Species," 191–92.
49. Keeney, *The Botanizers*.
50. Shteir, *Cultivating Women, Cultivating Science*, 35–36.
51. Martyn, "Observations on the Language of Botany," 147–48.
52. Shteir, *Cultivating Women*, 156–57.
53. Stoler and Cooper, "Between Metropole and Colony," 28.
54. "International Botanical Congress, 3rd 1910 Circulaire" and "Congrès international de botanique Bruxelles 1910: 4eme circulaire: Sur la bibliographie et la documentation botaniques," International Botanical Congress, Vertical Files, LTML-NYBG.
55. Somsen, "A History of Universalism," 365–67.
56. *Oxford English Dictionary*, under "vernacular."
57. Noyes, *Humble Theory*, 61–62.
58. Taylor, "Characteristics of German Folklore Studies," 293–301; Thomas, *Orientalists, Propagandists, and Ilustrados*; Mojares, *Isabelo's Archive*; Shimamura, "What Is Vernacular Studies?"

59. Tilley, "Global Histories, Vernacular Science, and African Genealogies," 110–19.

60. Parreñas, "From Decolonial Indigenous Knowledges to Vernacular Ideas in Southeast Asia," 413–16.

61. Ogilvie, *The Science of Describing*, 36–37.

62. Ogilvie, *The Science of Describing*, 215–29.

63. Linnaeus, *Species plantarum*, iv.

64. Kaempfer, *Amoenitatum exoticarum politico-physico-medicarum fasciculi* V, 604; Denis Diderot et al., *Encycopédie* 17:649.

65. Linnaeus, *A System of Plants*, 13.

66. Martyn, *The Language of Botany*, xi; also referenced in Martyn, "Observations on the Language of Botany," 147–48.

67. Cañizares-Esguerra, *How to Write the History of the New World*, 283–84.

68. Cañizares-Esguerra, *Nature, Empire, and Nation*, 56.

69. "5th Circulaire of the Permanent Committee of the International Congress of Botany"; "International Botanical Congress (2nd: 1905—Vienna, Austria)"; "International Botanical Congress, 3rd 1910 Circulaire: Congrès international de botanique Bruxelles 1910: Première circulaire," International Botanical Congress, Vertical Files, LTML-NYBG.

70. Gordin, *Scientific Babel*, 48–49.

71. Joaquin, *A Question of Heroes*, 66–69.

72. Thomas, *Orientalists, Propagandists, and Ilustrados*, 47–96.

73. W. Anderson, "Introduction: Postcolonial Technoscience," 644.

74. *Oxford English Dictionary*, under "sovereign."

75. B. Anderson, *Imagined Communities*.

76. For a concise political and intellectual history of the term in Indigenous activism and scholarship, see Brown, "Sovereignty," 81–90. Other deployments include the special issue "Native Feminisms: Legacies, Interventions, and Indigenous Sovereignties," edited by Goeman and Denetdale; Mortimer, "Kateri's Bones," 55–86. Mortimer engages Alfred, *Peace, Power, Righteousness*. On language and sovereignty, see Viatori and Ushigua, "Speaking Sovereignty," 7–21. On self-determination vis-à-vis the national body politic, see Robbins, "A Nation Within?," 257–74. See also Rutherford's focus on some of Biaks' shared vision of a "full sovereignty" in *Laughing at Leviathan*, 3.

77. Nadasdy, *Sovereignty's Entailments*, 46. Nadasdy compellingly opens with Alfred, "Sovereignty," in Barker, *Sovereignty Matters*, 39.

78. Barker, "Sovereignty," in *Keywords for Gender and Sexuality Studies*. For more elaboration on "critical sovereignty" in Indigenous studies and its taking account of gender, sexuality, and feminism, see Barker, *Critically Sovereign*.

79. Blok, "War of the Whales," 60–70.

80. Somsen, "A History of Universalism," 363–70.

81. Rutherford, *Laughing at Leviathan*, 10–13.

82. I acknowledge the work of Manuela Lavinas Picq that offers the term "vernacular sovereignties" with respect to Indigenous women's roles in influencing international politics. For Picq, "Indigenous sovereignties are practiced in vernacular contexts, thus transforming a singular form of authority into a plurality of forms adaptable to contextual realities." Indigenous women strategically deploy international legal norms to, in effect, vernacularize conventional, state-based notions of sovereignty to create local enact-

ments of authority. Whereas Picq's important ethnographic study frames the vernacular "in opposition to the supranational, placing Indigenous political practices at the antipode of European ones," I see the vernacular in generative relation to that which is sovereign, particularly in the history of science. See Picq, *Vernacular Sovereignties*, 18–21. I laud Deborah A. Thomas and Joseph Masco's edited volume, *Sovereignty Unhinged*, for its examination of sovereignty through affect, eschewing the temporally linear, utilitarian manifestations of the concept. The organization of their volume by "opposite impulses," not unlike my making and *un*making, opens exciting theoretical possibilities for sovereignty within such tensions. See Thomas and Masco, *Sovereignty Unhinged*, 10.

83. Barad, *Meeting the Universe Halfway*, 214–17.

84. Nanda, "The Epistemic Charity of the Constructivist Critics of Science," 288.

85. Chambers and Gillespie, "Locality in the History of Science," 235. Chambers and Gillespie engage Nanda's writing and gently refute its position by pointing to the highly situated nature of science, as any other knowledge system.

86. Haraway, "Situated Knowledges," 575–99. The literature on epistemic violence is vast. Foundational resources for me include Spivak, "Can the Subaltern Speak?," 271–313; Santos, *Epistemologies of the South*.

87. Cañizares-Esguerra, *Nature, Empire, and Nation*, 13.

CHAPTER 1. AN ASYMPTOTIC TAXONOMY

1. Gutaker et al., "Genomic History and Ecology of the Geographic Spread of Rice," 492–98.

2. Junker, *Raiding, Trading, and Feasting*, 234–35.

3. Fernandez, *Palayok*, 15–16. See also Fernandez, "The World of Miguel Ruiz," 74–79.

4. Nabhan and St. Antoine, "The Loss of Floral and Faunal Story," 231–32.

5. Reedy et al., "A Mouthful of Diversity," 12–13.

6. Mears, *Rice Economy of the Philippines*, 7.

7. Aguilar, "Rice and Magic," 304–9.

8. Efforts dotted the late eighteenth century, most notably under Juan José Ruperto de Cuéllar y Villanueba (ca. 1739–1801) of the Real Compañía de Filipinas, founded in 1785 to oversee the Spanish monopoly of the colony. Though blueprints and instructions for the garden's development had been drafted, the project was abandoned, and Cuéllar lost his post at the Real Compañía. See Llanos, *Ang Pagbubukid ng Kalikasang*, 348–53. In 1821, Madrid sent a royal order to Governor-General Mariano Fernández de Folgueras (1766–1823) to establish a public botanical garden in Manila to facilitate the study of commercial and medicinal plants. Only two years later, Governor-General Juan Antonio Martínez (1769–1826) issued a similar order on behalf of a garden that could cultivate indigenous and exotic flora. Neither attempt saw progress. See Rodriguez, "El Jardín Botánico de Manila," 77–78.

9. Borromeo-Buehler, "The 'Inquilinos' of Cavite," 69–98; De Jesus, *Tobacco Monopoly in the Philippines*; Legarda, *After the Galleons*, and Legarda, "Economic Background of Rizal's Time."

10. Lopez, *Orígen é historia del Jardín Botánico*, 3–9.

11. Ultramar, legajo 527, expediente 1, número 2, AHN.

12. Sánchez, "La etnografía de Filipinas," 157–85.

13. Ultramar, leg. 527, exp. 1, núms. 19, 32, and 34, AHN.
14. Ultramar, leg. 527, exp. 1, núm. 37, AHN.
15. Ultramar, leg. 527, exp. 2, núm. 24, AHN; Ultramar, leg. 527, exp. 2, núm. 37, AHN.
16. Pardo de Figueroa, *Algunos escritos del Teniente*, 165; Rizal, *Noli me tangere*, 51.
17. Ultramar, leg. 527, exp. 2, núm. 24, AHN.
18. Ultramar, leg. 527, exp. 2, núm. 25, AHN. Emphasis in the original.
19. Ultramar, leg. 527, exp. 1, núm. 7, AHN.
20. A thorough example of this impact appears in the history of the botanical garden in Havana as investigated in Aguilera-Manzano, "Havana's Botanical Garden in the Construction of Cuban National Identity."
21. For a history of this shift, see Fradera, *Colonias para después de un imperio*.
22. Schmidt–Nowara, *The Conquest of History*, 166.
23. Schmidt–Nowara, 166; Junco, "The Formation of Spanish Identity," 22; Elizalde, "Imperial Transition in the Philippines," 158.
24. Llorca, "La historia natural en la España," 132. Studies of natural history of the early and late Spanish Empire include Cañizares-Esguerra, *Nature, Empire and Nation*; Barrera-Osorio, *Experiencing Nature*; De Vos, "Research, Development, and Empire," 55–79.
25. Hoeg, "The Reception of Charles Darwin in Spain," 141.
26. Llorca, "La Historia Natural en la España," 130.
27. Rafael, *Contracting Colonialism*, 26; R. Rodríguez, "Lexicography in the Philippines (1600–1800)."
28. Jorge Cañizares-Esguerra, personal communication, April 7, 2023.
29. Of the several works of missionary-naturalists, Juan José Delgado's (1697–1755) manuscript published as *Historia general sacro-profana, política, y natural de las islas poniente llamadas Filipinas* in Manila in 1892 lists ninety-three different rice varieties. See Diaz-Trechuelo, "Eighteenth Century Philippine Economy: Agriculture," 125.
30. Blanco, *Flora de Filipinas*, 274.
31. Mentrida, *Diccionario de la lengua Bisaya*, 143.
32. Blanco, *Flora de Filipinas*, 275.
33. Noceda and Sanlucar, *Vocabulario de la lengua tagala*, 85.
34. Blanco, *Flora de Filipinas*, 274–75.
35. Noceda and Sanlucar, *Vocabulario*, 129.
36. *Census of the Philippine Islands*, 88.
37. Aguilar, "Rice and Magic," 305.
38. Mears, *Rice Economy*, 8.
39. Aguilar, "Rice and Magic," 301–3. Several precolonial communities had relied on female labor for stages of rice agriculture, food preparation, and consumption practices. In the Visayas, early chroniclers noted how women usually had harvested rice with a customary amount reaped at a particular time of day. See Scott, "Sixteenth-Century Visayan Food and Farming," 294.
40. Blanco, *Flora de Filipinas*, 276.
41. Marian Pastor Roces notes the historical significance of "pounding processes" to aspects of music, oral traditions, pharmacology, and weaving in the Philippines. Visual

representations of unhusking, therefore, can lead to a number of cultural associations depending on context. See Pastor-Roces, Baldovino, and Tysmans, *Sinaunang Habi*, 21.

42. *Census of the Philippine Islands*, 87. From 1855 to 1857, the import quantity leapt from 4,505 kilograms to 2,824,867.

43. Aguilar, "Rice and Magic," 304–9.

44. Doeppers, *Feeding Manila in Peace and War, 1850–1945*, 15.

45. Doeppers, 19–46.

46. Ultramar, leg. 527, exp. 2, núms. 13 and 15, AHN.

47. Ultramar 527, exp. 2, núm. 8, pp. 45–49, AHN.

48. Ultramar, leg. 527, exp. 1, núms. 67 [listed as 57] and 69 [listed as 59], AHN.

49. Ultramar, leg. 527, exp. 2, núm. 94, AHN.

50. R. Reyes, "Collecting and the Pursuit of Scientific Accuracy," 77–78. Reyes engages with Cañizares-Esguerra, *Nature, Empire and Nation*.

51. Llorca, "La Historia Natural en la España," 118–19.

52. Bankoff, "The Science of Nature and the Nature of Science," 95.

53. Ultramar, leg. 527, exp. 2, núms. 7 and 27, AHN.

54. A dedicated plot for the cultivation of palay, or unhusked rice, at the JBM appears with greater detail in the mid-1870s records. See Ultramar 527, exp. 2, núm. 60, AHN.

55. Clausen, "On the Use of the Terms," 157.

56. *Census of the Philippine Islands*, 94–95. As of this writing, no other archival material of the JBM that describes the varieties in great detail has been found. More is yet to be discovered, though I presume much was lost to an 1897 fire that destroyed the JBM buildings.

57. Blanco, *Flora de Filipinas*, 282.

58. J. Reyes, "¿Quién fué Don Regino García y Baza?," 8. Unpublished manuscript, Manila, 1940, HL-UST.

59. Ultramar, leg. 527, exp. 2, núm. 8, p. 44, AHN.

60. Diaz-Trechuelo, "Eighteenth Century Philippine Economy," 125; Tremml-Werner, *Spain, China, and Japan in Manila, 1571–1644*, 141–58.

61. Ultramar, leg. 429, exp. 11, AHN.

62. Pérez and Blanco, "La Ceres española y la Ceres europea," 381–83.

63. Clausen, "On the Use of the Terms," 158.

64. For a thorough intellectual history on the topic, see Wilkins, *Species: A History of the Idea*. For a sense of twentieth- and twenty-first-century philosophical debates on the "species problem," some representative readings include Ghiselin, "A Radical Solution to the Species Problem," 536–44; Wilson, Barker, and Brigandt, "When Traditional Essentialism Fails"; Mishler, "Species Are Not Uniquely Real Biological Entities."

65. Linnaeus, *Linnaeus' Philosophia botanica*, 114–15, 257. Linnaeus's definition of variety does not change considerably by the 1753 publication of *Species plantarum*. Clausen, "On the Use of the Terms," 159–259.

66. Schiebinger, *Plants and Empire*, 196.

67. Adanson, *Familles des plantes*, cxv.

68. Wilkins, *Species*, 80. Wilkins cites Stafleu, "Adanson and the 'Familles des plantes.'"

69. Chater, Brummitt, Persoon, "Subspecies," 143–48.

70. Persoon, *Synopsis methodica fungorum*, 70.

71. Chater, Brummitt, Persoon, "Subspecies," 145. Emphasis mine.

72. Clausen, "On the Use of the Terms," 160.

73. Chater, Brummitt, Persoon, "Subspecies," 148.

74. Subramaniam, *Ghost Stories for Darwin*, 51.

75. Darwin, *On the Origin of Species*, 61–62.

76. Gray, "XXV.—Do Varieties Wear Out or Tend to Wear Out?," 174.

77. Asa Gray to Joseph Dalton Hooker, September 3, 1883, in Gray, *Letters of Asa Gray*, 744.

78. As was the case for the American Ornithologists' Union, known formerly as the Committee on Classification and Nomenclature (founded 1883). See Lewis, *The Feathery Tribe*, 159–80.

79. Darwin himself had an extensive Anglo-European correspondence network that included naturalists in the United States. A recent digital initiative, Darwin Correspondence Project, covers this well. See "Darwin Correspondence Project." See also McCoy, "Fatal Florescence," 16. McCoy references the 1885 Conference of Berlin and the Versailles Peace Conference of 1919, for example, as key political formations of the time.

80. Mazower, "The Strange Triumph of Human Rights, 1933–1950," 381.

81. Rodogno, *Against Massacre*, 4–12.

82. Clausen, "On the Use of the Terms," 161.

83. Candolle, *Laws of Botanical Nomenclature*, 39–41.

84. Nicolson, "A History of Botanical Nomenclature," 33–34.

85. Candolle, *Laws of Botanical Nomenclature*, 8–25.

86. Hoeg, "The Reception of Charles Darwin," 140.

87. "Listes des members du congrès," p. 7, ca. 1910, International Botanical Congress–Paris, 1900, LTML-NYBG.

88. Colmeiro, *Enumeración y revisión de las plantas de la Península Hispano-Lusitana é Islas Baleares*, vi–viii.

89. Subramaniam, *Ghost Stories for Darwin*, 7–14.

90. Linnaeus, *Linnaeus' Philosophia botanica*, 221.

91. Phillips, "The Taste Machine," 462–63.

92. Mandelkern, "Taste-Based Medicine," 11–16.

93. Otálora-Luna and Aldana, "The Beauty of Sensory Ecology," 20.

94. Noceda and Sanlucar, *Vocabulario*, 32–73.

95. Mentrida, *Diccionario*, 342.

96. Noceda and Sanlucar, *Vocabulario*, 66–68.

97. Blanco, *Flora de Filipinas*, 276; Noceda and Sanlucar, *Vocabulario*, 406.

98. Lisboa, *Vocabulario de la lengua Bicol*, 81. The original dictionary uses "rubia" to describe the husk color, though the Tagalog term generally refers to copper or red. Golden may be a color approximating the original description, though this cannot be fully determined. Binayoyo bears similarity to the Tagalog "binayo" to mean pounded using a mortar and pestle.

99. Noceda and Sanlucar, *Vocabulario*, 38–59.

100. Dupré, "Natural Kinds and Biological Taxa," 81–82.

101. *Exposición de Filipinas*, 218. I thank Michael Purugganan for enabling me to see García's preserved rice variety specimens at the Jardí Botànic de Barcelona, likely donated following the Barcelona exposition. Many of the varietal names (most hailing from Batangas) in Barcelona do not appear in García's JBM catalogs, indicating how much more expansive García's rice research had become by the late 1880s.
102. *Census of the Philippine Islands*, 88.
103. Lincoln Taiz, email communication with author, July 7, 2024.
104. Legarda, *After the Galleons*, 156–57.
105. Mears, *Rice Economy*, 8.
106. Legarda, *After the Galleons*, 166–67.
107. Doeppers, *Feeding Manila in Peace and War*, 18–46.
108. Rodriguez, "El Jardín Botánico de Manila," 80.
109. Ultramar, leg. 527, exp. 2, núm. 31, AHN.

CHAPTER 2. SCIENTIFIC STATECRAFT

1. Ultramar, legajo 527, expediente 2, número 19, AHN.
2. Mojares, *Brains of the Nation*, 451.
3. Studies of science and imperial statecraft are abundant and include: MacLeod, "On Visiting the 'Moving Metropolis'"; Gascoigne, *Science in the Service of Empire*; Yalçinkaya, *Learned Patriots*; Philip, *Civilizing Natures*.
4. Chakrabarti and Worboys, "Science and Imperialism since 1870," 14–20.
5. Qureshi, "Science, Empire and Globalization in the Nineteenth Century."
6. J. Reyes, "¿Quién fué Don Regino García y Baza?," 28, HL-UST. Following Vidal's death, García described Vidal as "my distinguished companion." Regino García to Robert Allen Rolfe, May 8, 1890, folio 288, Directors' Correspondence 165/288, Royal Botanic Gardens, Kew.
7. US botanist Elmer D. Merrill (1876–1956), whom I discuss in later chapters, disregarded most Spanish-era botanical production save for Vidal's, arguably elevating Vidal's work to prominence in Anglophone histories of the science in the Philippines. Among many of Merrill's writings on Vidal, *Botanical Work in the Philippines* is an exemplary assessment.
8. Merrill, *A Discussion and Bibliography of Flowering Plants*, 50–52.
9. Van Steenis, *Flora Malesiana*, xi.
10. Stoler and Cooper, "Between Metropole and Colony," 28.
11. Vidal, *Memoria*, 5.
12. Bankoff, "The Tree as the Enemy of Man," 321–28.
13. Ultramar, leg. 526, exp. 8, núm. 1, AHN.
14. Orillos-Juan, "Ang Inspección General de Montes," 80.
15. Jordana y Morera, *Memoria sobre la producción de los montes públicos de Filipinas*, 5–6.
16. Burzynski, "The Timber Trade and the Growth of Manila, 1864–1881," 168–69.
17. Ultramar, leg. 526, exp. 8, núm. 57, AHN.
18. Ultramar, leg. 528, exp. 1, núm. 1, AHN.
19. Ultramar, leg. 528, exp. 1, núm. 4, AHN.

20. De Caleya, "Sebastián Vidal y Soler."
21. Vidal, *Memoria*, 18–19; 217.
22. Rizal reportedly found Vidal's *Reisen* translation subpar and believed he himself could have produced better. See Mojares, "Jose Rizal in the World of German Anthropology," 173.
23. Jagor, *Viajes por Filipinas*, vii–viii.
24. Weston, *Specters of Germany*, 16–17.
25. José Malcampo y Monge to the Sr. Ministro de Ultramar, September 27, 1875, Varios Personajes, 21388–21389, NAP.
26. "Bibliografia filipina," *Revista de Filipinas*, Tomo 1, July 1875–June 1876, núm. 16, 471, February 1, 1876, HL-UST; "Memoria leida por el director de la Escuela practica professional de artes y oficios," 1891, Memorias, NAP.
27. Weston, *Specters of Germany*, 62–63. For example, Vidal's *Sinopsis* relied considerably on German botanist Wilhelm Sulpiz Kurz's *Forest Flora of British Burma* (1877).
28. Raes and van Welzen, "The Demarcation and Internal Division of Flora Malesiana: 1857–Present," 6–7.
29. Weston, *Specters of Empire*, 6.
30. Ultramar, leg. 524, exp. 14, núm. 32, AHN.
31. Ultramar, leg. 524, exp. 14, núm. 28, AHN.
32. Vidal, *Memoria*, 113–36.
33. *Gaceta de Madrid*, 276, contained in Ultramar, leg. 534, exp. 2, núm. 7, AHN.
34. Ultramar, leg. 524, exp. 26, AHN.
35. Ultramar, leg. 524, exp. 14, núm. 30, AHN.
36. Sánchez, *Un imperio en la vitrina*, 98.
37. *Gaceta de Madrid*, 277, contained in Ultramar, leg. 534, exp. 2, núm. 7, AHN.
38. Ultramar, leg. 534, exp. 2, núm. 56, AHN.
39. Ultramar, leg. 534, exp. 3, núm. 2, AHN.
40. Ultramar, leg. 534, exp. 2, núm. 57, AHN.
41. Santiago, "The Painters of *Flora de Filipinas* (1877–1883)," 93.
42. Parco-De Castro, "Fostering Social Transformation through Philippine Secondary Education," 49.
43. Ultramar, leg. 527, exp. 1, núm. 73 (listed as 63 in the AHN).
44. Luciano P. R. Santiago, "Philippine Academic Art: The Second Phase (1845–98)," 80. Juan Luna y Novicio (1857–1899) also studied at the school. Mathematics courses had been allegedly very difficult at the Escuela Náutica, and dropouts were common. See Frederick Fox, "Philippine Vocational Education: 1860–1898," 267.
45. Santiago, "Damian Domingo and the First Philippine Art Academy (1821–1834)," 272; Quirino, "Damián Domingo, Filipino Painter," 79.
46. On the discrepancies behind the ADP's establishment, see Flores, *Painting History*, 202–14.
47. Santiago, "Philippine Academic Art," 67.
48. Quirino, "Manila's School of Painting," 348.
49. Díaz Pascual, "Modelos para la escuela de pintura de Manila."
50. Flores, *Painting History*, 77–90. See also Maravall, *Culture of the Baroque*.
51. Jurilla, *Tagalog Bestsellers*, 31.
52. Flores, *Painting History*, 271–83.

53. Santiago, "Philippine Academic Art," 76–79.
54. Santiago, "Philippine Academic Art," 68. Santiago distinguishes "Philippine academic art" from "native art," which he defines as produced by those who learned by apprenticeship and by self-taught artists.
55. Santiago, "The Painters of *Flora de Filipinas* (1877–1883)," 98.
56. Kroupa, "Georg Josef Kamel," 181.
57. Santiago, "The Painters of *Flora de Filipinas* (1877–1883)," 87.
58. Raquel A. G. Reyes, "Collecting and the Pursuit of Scientific Accuracy: The Malaspina Expedition in the Philippines, 1792," 77–8. Reyes engages with Jorge Cañizares-Esguerra, *Nature, Empire, and Nation*.
59. Llanos, *Ang Pagbubukid ng Kalikasang*, 351–53.
60. Bleichmar, *Visible Empire*, 125.
61. Santiago, "The Painters of *Flora de Filipinas* (1877–1883)," 88. See also Bañas Llanos for more extensive treatment of the illustrations from the Cuéllar expedition.
62. Reyes, "Collecting and the Pursuit," 78. See also Galera Gómez, *El arca de Née*.
63. Santiago, "The Painters of *Flora de Filipinas* (1877–1883)," 87–88.
64. Santiago, "Philippine Academic Art," 75.
65. Roxas, *The World of Felix Roxas*, 72–73. Roxas estimates Jackson's birth year as 1855. Other sources suggest 1858.
66. Gutierrez, "Emina María Jackson y Zaragoza," 242–46.
67. Bleichmar, *Visible Empire*, 53.
68. Bleichmar, *Visible Empire*, 63.
69. Daston, "Type Specimens and Scientific Memory," 160.
70. Daston and Galison, *Objectivity*, 19–27.
71. Ultramar, leg. 534, exp. 2, núm. 35, AHN.
72. Ultramar, leg. 534, exp. 2, núm. 23, AHN.
73. Sebastián Vidal y Soler, *Sinopsis de familias y generos*, vii–xi; Ultramar, leg. 534, exp. 3, núm. 2, AHN.
74. Ultramar, leg. 524, exp. 26, núm. 44, AHN.
75. Ultramar, leg. 534, exp. 2, núm. 58, AHN.
76. Netzorg, "Books for Children in the Philippines," 285.
77. Buhain, *A History of Publishing in the Philippines*, 15.
78. Ultramar, leg. 534, exp. 3, núm. 39, AHN.
79. Santiago, "The Painters of *Flora de Filipinas*," 87–92; Ultramar, leg. 534, exp. 2, núm. 58, AHN.
80. *Exposición de Filipinas*, 46.
81. Santiago, "The Painters of *Flora de Filipinas*," 94–96.
82. Flores, *Painting History*, 252.
83. Santiago, "Philippine Academic Art," 72.
84. Bleichmar, *Visible Empire*, 82.
85. Santiago, "Philippine Academic Art," 68–80.
86. Bleichmar, *Visible Empire*, 152–54. I separately apply Bleichmar's astute notion of pictorial "white space" to US botany during the Philippine Commonwealth period. See Gutierrez, "*Cycas wadei* and Enduring White Space."
87. Bleichmar, *Visible Empire*, 84–122.

88. Bleichmar, *Visible Empire*, 151.

89. Capistrano-Baker, "Whither Art History?," 248.

90. My interpretation maintains some affinity with Bleichmar's interpretation of Spanish American artists of the eighteenth century, whose artistic depictions of the environment departed from European natural history conventions. Their works, though catering to foreign European audiences, "provide highly constructed, selective, and stylized visions that adhere to their own pictorial conventions." See Bleichmar, *Visible Empire*, 161–84.

91. Nick Joaquin romanticized the nineteenth-century "Filipino" artist, having written, "And what's why I would say that the Filipino as painter was the First Filipino. He was the first to exemplify, both in himself and in his works, those virtues that were said to be lacking in the Indio; and by making those virtues define the identity of the Filipino, he advanced the process of our evolution from Indio to Filipino." See Joaquin, introduction to *Juan Luna, the Filipino as Painter*. See also Flores's interpretation of Joaquin's assertion vis-à-vis the ilustrado monopoly over nationalist production, in *Painting History*, 27–29.

92. *Exposición de Filipinas*, 46.

93. Ultramar, leg. 528, exp. 3, núm. 2, AHN.

94. Ultramar, leg. 534, exp. 2, núm. 34, AHN.

95. Bleichmar, *Visible Empire*, 141–47.

96. Barnard, *Nature's Colony*, 7.

97. Ultramar, leg. 528, exp. 3, núm. 3, AHN.

98. Ultramar, leg. 528, exp. 3, núm. 4, AHN.

99. Santiago, "Philippine Academic Art," 75.

100. Pilapil, "The Cause of the Philippine Revolution," 255–56.

101. Claudio, *Jose Rizal*, 10–11.

102. Ultramar, leg. 534, exp. 3, núm. 29, AHN.

103. Ultramar, leg. 524, exp. 14, núm. 44, AHN.

104. Ultramar, leg. 524, exp. 14, núm. 45, AHN.

105. Ultramar, leg. 528, exp. 3, núm. 5, AHN.

106. *Presupuestos generales de gastos é ingresos*, 46–49.

107. Ultramar, leg. 524, exp. 14, núm. 62, AHN.

108. Ultramar, leg. 524, exp. 14, núm. 58, AHN.

109. Pasaporte 21195–21196, Varios Personajes, NAP.

110. Jaena, "Filipinas en la Exposición universal de Barcelona," 247.

111. Schmidt-Nowara, *The Conquest of History*, 167–69.

112. Ultramar, leg. 524, exp. 14, núm. 62, AHN.

113. Scott, *History on the Cordillera*, 13; Aguilar, "Tracing Origins," 614–16.

114. Aguilar, "Tracing Origins," 614.

115. Aguilar, "Tracing Origins," 615.

116. Aguilar, "Tracing Origins," 615–16.

117. For studies of the Exposición in Madrid and the Philippine exhibits at the 1888 Barcelona Universal Exposition, see Sánchez, "Indigenous art at the Philippine Exposition of 1887," 283–94; Aguilar, "Romancing Tropicality 'Ilustrado' Portraits of the Climate in the Late Nineteenth Century"; Flores et al., "Polytropic Philippine."

118. *Exposición de Filipinas*, 46.
119. Hernández and de Piquer, *Crónica de la Exposición de Filipinas*, 63–73.
120. *Exposición de Filipinas*, 46–47.
121. Ultramar, leg. 524, exp. 14, núm. 68, AHN.
122. J. Reyes, "¿Quién fué Don Regino García y Baza?," 31–33. García's compiled works include *Catálogo de las plantas del herbario recolectado por el personal de la suprimida Comisión de la Flora Forestal* (1892). The *Catálogo* does not list an author and is attributed to the Comisión. The Santo Tomas press published the *Catálogo*, which intimates García may have been behind it.

CHAPTER 3. UBIQUITOUS SAMPAGUITA

1. Dery, 91–93. I have translated "Jocelynang Baliwag" for the epigraph.
2. Molina, *Ang Kundiman ng Himagsikan*, 15–29.
3. De León, "Poetry, Music and Social Consciousness," 271–72; Castro, *Musical Renderings of the Philippine Nation*, 183–84; Lumbera, *Pag-akda ng bansa*, 98–99.
4. Reyes, *Love, Passion and Patriotism*, 178. See also Rafael, "Nationalism, Imagery and the Filipino Intelligentsia," 136–46.
5. Hau, *Necessary Fictions*, 50–51.
6. Planta, *Traditional Medicine in the Colonial Philippines*, 38. Planta cites Panganiban, *Diksyunaryo-tesauro Pilipino-Ingles*.
7. Lim, *Edible Medicinal and Non-Medicinal Plants*, vol. 8, 530.
8. Aquino, "Veto Message of President Aquino on House Bill No. 5655"; Cepeda, "House Pushes Bills Declaring Waling-Waling, Balangay as National Symbols"; Congress of the Philippines, Senate Bill No. 3307, An Act Declaring the Waling-Waling Orchids as National Flower of the Philippines in addition to Sampaguita, October 16, 2012, https://www.senate.gov.ph/lisdata/1423611973!.pdf.
9. Besky and Padwe, "Placing Plants in Territory," 10. See also Cañizares-Esguerra's discussion of the ahuehuete (or Mexican cypress, *Taxodium mucronatum*) within a "historically informed type of landscape aesthetics" in nineteenth-century Mexico, *Nature, Empire, and Nation*, 136–39; Choy's treatment of the *Spiranthes hongkongensis* in *Ecologies of Comparison*, 53–72; the fynbos (or fine bush) and the flower that most emblematizes it (*Protea cynaroides*) in the modern heritage of the Western Cape and more broadly, South Africa, in Comaroff and Comaroff, "Naturing the Nation," 627–51; in the contemporary debate over "native" versus " invasive" species discourse captured in Mastnak, Elyachar, and Boellstorff, "Botanical Decolonization," 363–80; and Braverman's coverage of pine forests and olive groves in the contestations over landscape between Israelis and Palestinians in *Planted Flags*.
10. Jurilla, *Tagalog Bestsellers*, 29.
11. Santiago, "The Flowering Pen," 570–71.
12. Mojares, *Isabelo's Archive*, 214–16.
13. Aguilar, "Tracing Origins," 605–37; Thomas, *Orientalists, Propagandists, and Ilustrados*; Mojares, "Jose Rizal in the World of German Anthropology," 163–94; see also Mojares, *Isabelo's Archive*; Mojares, *Brains of the Nation*.
14. Reyes, "¿Quién fué Don Regino García y Baza?," 34.

15. Welch, "Atrocities in the Philippines," 234; Santiago, "Philippine Academic Art," 84. Santiago cites *The World of Felix Roxas*, 72, but this does not correspond to García's alleged signing of the constitution.

16. Roberts, "U.S. Forestry in the Philippines," 318.

17. Mojares, *Brains of the Nation*, 499–500.

18. Bleichmar, "Visible Empire," 450.

19. Rafael, *Contracting Colonialism*, 19. Rafael further cites Schumacher, "The Manila Synodal Tradition: A Brief History," 309, and Blair and Robertson, *The Philippine Islands*, vol. 20:250–52.

20. Mojares, *Brains of the Nation*, 384.

21. Alcina, *History of the Bisayan People*, 15.

22. Barad, *Meeting the Universe Halfway*, 128; Shapin, "Pump and Circumstance," 481.

23. Alcina, *History of the Bisayan People*, 578.

24. Rafael, *Contracting Colonialisms*, 115–21.

25. Noceda and Sanlucar, *Vocabulario de la lengua tagala*, 283.

26. Alcina, *History of the Bisayan People*, 15.

27. Lumbera, *Tagalog Poetry, 1570–1898*, 8–9. I have very slightly adjusted Lumbera's translation of the proverb from Tagalog and Spanish, though the overall meaning and impact remain.

28. Noceda and Sanlucar, *Vocabulario*, 283.

29. Kroupa, "*Ex epistulis Philippinensibus*," 229–59.

30. Kamel, "Tractatus de plantis Philippinensibus scandentibus," 1838–40.

31. Kroupa, email communication with author, April 10, 2020.

32. Kroupa, "Georg Joseph Kamel," 181.

33. Blanco, *Flora de Filipinas*, 9.

34. Linnaeus, *Species plantarum*, 6. Linnaeus cited *Hortus cliffortianus* (1737) and *Flora leydensis* (1740), as well as his own *Hortus upsaliensis* (1748) and *Flora zeylanica* (1747), which include information on the *Nycanthes* but do not cite the species *N. sambac*. For Monandria, see Linnaeus, *Systema naturae*, 847.

35. Rumphius, *Herbarium amboinense*, vol. 5, 52–53.

36. Van Rheede, *Hortus indicus malabaricus*, vol. 6, 89. While in the Philippines, Spanish naturalist Juan de Cuéllar compared specimens to those found in van Rheede's flora. Cuéllar equated the "sambaga" (from the Spanish language, in his annotations) or Tagalog "capopot" to the "nalla nulla" and "cudda mulla" identified by van Rheede. Cuéllar would have made the identification sometime during or after 1795 once he received van Rheede's multivolume work in Manila. It is unclear if he cross-referenced from *Hortus kewensis*, which was published some six years before. See Bañas Llanos, *Ang Pagbubukid ng Kalikasang*, 331–36.

37. "Distribution géographique des récoltes de *Jasminum sambac (C. Linnaeus) D. C. Solander x W. Aiton*," Les taxons, Nadeaud Database of the Herbarium of French Polynesia, n.d. Accessed July 10, 2020. http://nadeaud.ilm.pf/details-taxon/2885; Aiton, *Hortus Kewensis*, 8.

38. "Nyctanthes sambac L.," n.d., Linnaean herbarium (S-LINN), Department of Phanerogamic Botany, Swedish Museum of Natural History. Accessed April 10, 2020. http://linnaeus.nrm.se/botany/fbo/n/bilder/nycta/nyctsam1.jpg. The *Nyctanthes sambac* L. has been preserved in the Linnaean herbarium, which cites India as the *N. sambac* habitat.

39. Blanco, *Flora de Filipinas*, 9.
40. Rumphius, *Herbarium amboinense*, 52.
41. Vidal y Soler, *Sinopsis de familias y generos*, 180–81.
42. Ultramar, leg. 534, exp. 2, núm. 35, AHN.
43. Rafael, "The War of Translation," 297.
44. Lumbera, *Tagalog Poetry, 1570–1898*, 87–136.
45. Lumbera, *Tagalog Poetry, 1570–1898*, 138–40.
46. Camacho, *100 Taon*, 117.
47. Brillantes, "La Flor de Manila," 235.
48. Real Academia Española, *Diccionario de la Lengua Española*, under "gentil," accessed August 30, 2024, https://dle.rae.es/gentil.
49. Real Academia Española, *Diccionario de la Lengua Española*, under "peregrina," accessed August 30, 2024, https://dle.rae.es/peregrina.
50. This is clear in composer Levi Celerio's Tagalog translation of the song, which adheres firmly to the gentil-collective notion of the sampaguita: "Sampaguita of our people / flower of extreme refinement / you are the chosen gem / as the symbol of our race" ("Sampaguita ng aming lipi / bulaklak na sakdal ng yumi / Ikaw ang mutyang pinili / na sagisag ng aming lahi." See Canonigo, *Tinig ng Bayan*, 54.
51. Santiago, "The Flowering Pen," 585.
52. Mojares, *Isabelo's Archive*, 7–12.
53. "Rimmang-ayca á rimmangpaya / iti asi quen dungngo ti Ama quen Ina / balasangcan Mellang, nğa aoan ti curangna, / á cas agucrad á sampaga." De los Reyes, *El Folk-lore filipino*, 194.
54. King, *Bloom*, 12. See also Müller-Wille, "Linnaeus and the Love Lives of Plants," and Taiz and Taiz, *Flora Unveiled*.
55. Scott, "Class Structure in the Unhispanized Philippines," 139–59.
56. Lumbera, *Tagalog Poetry, 1570–1898*, 13.
57. De los Reyes, *El Folk-lore filipino*.
58. Laktaw, *Diccionario Hispano-Tagalog*, 321.
59. Strikingly, the "jazmín" remains untranslated in contemporary Philippine cultural production as in Ryan Cayabyab's "Tunay Na Ligaya" (True joy) released in 1999: "I do not notice the scent of jasmine when I am with you, darling." Or in the song "Sa Piling ng Bulaklakan" (Among the Flowers) that differentiates the sampaguita from the "hasmin," though both are the "color of morning" and as "delicious as hope."
60. De los Reyes, *El Folk-lore filipino*, 192.
61. Mojares, *Brains of the Nation*, 179.
62. De los Reyes, *El Folk-lore filipino*, 157–58.
63. Mojares, *Isabelo's Archive*, 3.
64. Thomas, *Orientalists, Propagandists, and Ilustrados*, 106.
65. Rizal, "De Rizal a Barrantes—Réplica de Rizal a la crítica de Barrantes contra el 'Noli,'" in *Epistolario Rizalino, 1887–1890*, vol. 2, 301–2; Pardo de Tavera, *Plantas medicinales de Filipinas*, 304; Jaena, "Filipinas en la Exposición Universal de Barcelona," 247.
66. Pardo de Tavera, *Plantas medicinales de Filipinas*, 22–36.
67. Pardo de Tavera, *Plantas medicinales de Filipinas*, 192–93.
68. Berlin, Breedlove, and Raven, "Covert Categories and Folk Taxonomies," 290–99.

69. Pardo de Tavera, *Plantas medicinales*, 10.

70. Pardo de Tavera, *El sanscrito en la lengua tagalog*, 48. There is contemporary agreement with Pardo's claims. See Planta, *Traditional Medicine in the Colonial Philippines*, 38; "*Jasminum sambac*," Missouri Botanical Garden, accessed April 8, 2020. http://www.missouribotanicalgarden.org/PlantFinder/PlantFinderDetails.aspx?taxonid =282952&isprofile=1&basic=Jasminum%20sambac.

71. Santiago, "The Painters of *Flora de Filipinas*," 92–93.

72. Thomas, "Isabelo de los Reyes and the Philippine Contemporaries of *La Solidaridad*," 390. The serial also featured colonial botany work of the time. García appeared on the cover of issue number 82 in 1893; Vidal on that of issue number 31 in 1892.

73. Clark, "The Worlding of the Asian Modern," 74; Santiago, "Damian Domingo and the First Philippine Art Academy," 278.

74. Roces, "Costumes, colleciónes del trajes, costumbres, costumbrista"; see also Blanco, "Aesthetics" in *Frontier Constitutions*, 157–83.

75. A vast literature exists on this topic, and especially instructive are Yuval-Davis, *Gender and Nation*; Das, *Life and Words*; Alexander, "M/othering the Nation"; Thapar-Björket, "Gender, Nations and Nationalism," 803–33.

76. On the "Modern Woman" in ilustrado work, namely that by Juan Luna, see Reyes, *Love, Passion and Patriotism*, 49. On the feminine ideal, Reyes writes, "The Filipina . . . was believed by the propagandistas to possess inherent qualities of tenderness, sweetness, gentleness, docility and passivity, attributes they saw as the appropriate, natural and biologically-ordained complement to the sexual identity of men." See xxix.

77. Only about 5 percent of the Philippine population could access Spanish by the turn of the century. Rafael, "The War of Translation," 284.

78. Agoncillo, *Malolos*, 145.

79. Mojares, *Brains of the Nation*, 20–21.

80. Pardo de Tavera, *Medicinal Plants of the Philippines*, 158. Emphasis mine.

81. Mojares, "The Formation of Filipino Nationality under U.S. Colonial Rule," 11–32.

82. Abinales, "American Rule and the Formation of Filipino 'Colonial Nationalism,'" 604–21.

83. Mojares, *Isabelo's Archive*, 245.

84. "Our State Flowers: The Floral Emblems Chosen by the Commonwealths," 481–517; "The National Floral Emblem."

85. Gibbs, *Advanced English Grammar and Composition*, 323. Emphasis original. The ilang-ilang had also been an important adornment in places beyond Manila, including Mindanao.

86. Chavez, "The Sampaguita," 13.

87. Rodriguez, *The Legend of the Sampaguita*.

CHAPTER 4. WOVEN TRANSFORMATIONS

An earlier version of portions of this chapter appeared in *History Compass* as "From Objects of Study to Worldmaking Beings: The History of Botany at the Corner of the Plant Turn." I thank the three reviewers with *History Compass* for recognizing the potential in plant worldmaking.

1. "Textile for the panapisan," 70.1/5429A, Laura Watson Benedict Collection, Division of Anthropology, AMNH; Benedict, *A Study of Bagobo Ceremonial, Magic and Myth*, 209–10.

2. Roces, Baldovino, and Tysmans, *Sinaunang Habi*, 16.

3. Agoncillo, *Prelude to 1896*, 2. See also Agoncillo, *Introduction to Filipino History*, 59–60; Planta, *Traditional Medicine in the Colonial Philippines*, 79; Zaide, *History of the Filipino People*, 220–21.

4. Quizon, "Between the Field and the Museum," 292.

5. The other extensive collections include that gathered by Fay-Cooper Cole (1881–1961) in Chicago's Field Museum and by the sisters Elizabeth and Sarah H. Metcalf at the Penn Museum in Philadelphia and the Smithsonian Institution in Washington, DC. See Quizon, 290–91. In this chapter, I also reference holdings in the Metcalf collection of the Smithsonian, which bear processual similarity to the Benedict collection. However, Benedict's collection is the principal focus of my analysis.

6. Thorough studies of the exposition include Rydell, *All the World's a Fair*; Fermin, *1904 World's Fair*; Grindstaff, "Creating Identity," 245–63; Burns, "'Which Way to the Philippines?,'" 21–48.

7. Quizon, "Between the Field and the Museum," 290–91.

8. Subramaniam, *Ghost Stories for Darwin*, 27.

9. Barad, *Meeting the Universe Halfway*, 90.

10. Gutierrez, "Botanical Knowledge within Itneg Weaving and Dyeing," 71–78.

11. Roces, Baldovino, and Tysmans, *Sinaunang Habi*, 16.

12. Bsumek, *Indian-Made*, 4. See also 143–73.

13. Roces et al., *Sinaunang Habi*, 11.

14. Klock, "Agricultural and Forest Policies of the American Colonial Regime," 3–19; Ventura, "From Small Farms to Progressive Plantations," 459–83. On US incapacity to develop a flourishing timber industry, see Luyt, "Empire Forestry and Its Failure in the Philippines, 1901–1941," 66–87; N. Roberts, "U.S. Forestry in the Philippines."

15. Fradera, "Reading Imperial Transitions," 39–41.

16. McCormick, "From Old Empire to New," 64–66.

17. C. Johnson, "'Alliance Imperialism,'" 122–26.

18. McCormick, "From Old Empire to New," 68–69.

19. McCoy et al., "On the Tropic of Cancer," 5.

20. W. Anderson, "Science in the Philippines," 299. Anderson cites Stanley, *A Nation in the Making*, and Owen, *Compadre Colonialism*.

21. Overfield, *Science with Practice*, 74; Smocovitis, "One Hundred Years of American Botany," 942.

22. Mickulas, *Britton's Botanical Empire*, 158.

23. For the full law, see United States Philippine Commission, An Act Providing for the Establishment of Government Laboratories for the Philippine Islands. See also W. Anderson, "Science in the Philippines," 209.

24. Planta, *Traditional Medicine in the Colonial Philippines*, 106.

25. "Laboratories of the Bureau of Science, Philippine Government," *Far Eastern Review*, 148, Periodicals, AHC.

26. Freer, "The Work of the Bureau of Government Laboratories, of the Philippine Islands," 108.

27. W. Anderson, "Science in the Philippines," 300.
28. "Laboratories of the Bureau of Science," 148.
29. US Bureau of Insular Affairs, *Official Handbook, Description of the Philippines*, Part I, 77.
30. Brown, preface to *Minor Products of Philippine Forests*, 3.
31. Brown and Fischer, "Philippine Forest Products as Sources of Paper Pulp," 3.
32. Brown and Merrill, "Philippine Fiber Plants," 323.
33. Gutierrez, "Botanical Knowledge within Itneg Weaving and Dyeing," 67–88.
34. Watson-Verran and Turnbull, "Science and Other Indigenous Knowledge Systems"; Gutierrez, "Rehabilitating Botany in the Postwar Moment," 54–55.
35. Drayton, *Nature's Government*, 55–67.
36. Of the generous amount of literature on these topics, see Osborne, "Acclimatizing the World," 135–51; De Vos, "The Science of Spices," 399–427; De Vos, "An Herbal El Dorado," 117–21; Schiebinger, *Plants and Empire*, 35–44.
37. Schiebinger, *Plants and Empire*, 11.
38. Dacudao, *Abaca Frontier*, 223–37.
39. Stanley, "Commercial Conditions in Panay and Negros," *Manila Daily Bulletin*, 27, Periodicals, AHC.
40. N. Roberts, "U.S. Forestry in the Philippines," 117.
41. Hanson, "The Great Hemp District of Davao," *Manila Times*, 108, Periodicals, AHC.
42. US Bureau of Insular Affairs, *Official Handbook, Description of the Philippines*, pt. 1, 79–80.
43. Reid, "Violence at Sea," 24. Reid cites Warren, *The Sulu Zone, 1768–1898*.
44. Tagliacozzo, *Secret Trades, Porous Borders*, 84–87.
45. Charbonneau, "'A New West in Mindanao,'" 304–23.
46. Paredes, "Experimental Science for the 'Bananapocalypse.'" Paredes engages Tiu, *Davao 1890–1910*; Vellema, Borras, and Lara, "The Agrarian Roots of Contemporary Violence Conflict in Mindanao, Southern Philippines," 298–320; and Charbonneau, "'A New West in Mindanao,'" 304–23.
47. Charbonneau, *Civilizational Imperatives*, 15. Charbonneau also cites Gowing, "Mandate in Moroland."
48. Hayase, "Tribes of the Davao Frontier," 139–40.
49. Merrill, "The Present Status of Botanical Exploration of the Philippines," 161.
50. Copeland, *Polypodiaceae of the Philippine Islands*, 144.
51. Dacudao, *Abaca Frontier*, 166ni.
52. Dacudao, *Abaca Frontier*, 244.
53. Hayase, "Tribes on the Davao Frontier, 1899–1941," 141–44; "Murdered by Natives," *Santa Barbara Morning Press*, 1906. See also Dacudao, *Abaca Frontier*, 166. There is some dispute about the precise spelling of Mungalayan's name.
54. Hayase, "Tribes on the Davao Frontier," 142–44; Quizon, "Two Yankee Women at the St. Louis Fair," 534–35. As Dacudao argues, the introduction of the pakyaw, a profit-sharing agreement between landowners and workers tied to the sale of abacá, may have turned the tide of conflict between tribal and foreign residents. Over decades after its implementation, instances of violence were comparatively less common. See Dacudao, *Abaca Frontier*, 168–72.

55. Quizon, "Between the Field and the Museum," 292.
56. Vidal y Soler, *Memoria sobre el ramo de montes en las islas Filipinas*, 194–202.
57. García y Baza, prologue to *Los arboles de goma, resinas y frutos oleosos de Filipinas*.
58. Merrill, "Report of Elmer D. Merrill, Botanist," 464.
59. Afable, "Journeys from Bontoc to the Western Fairs, 1904–1915," 446–47.
60. Rydell, *All the World's a Fair*, 329.
61. Bernstein, "The Perils of Laura Watson Benedict," 168–72.
62. Bsumek, *Indian-Made*, 150–52.
63. Quizon, "Between the Field and the Museum," 292.
64. Benedict, *A Study of Bagobo Ceremonial*, 66–70.
65. Benedict, *A Study of Bagobo Ceremonial*, 80–81.
66. Benedict, "Tolus Ka Talegit,'" manuscript, p. 3, Laura Watson Benedict Purchase, Archives of the Division of Anthropology, AMNH.
67. Coo, *Clothing the Colony*.
68. Spencer, "Abacá and the Philippines," 95.
69. Née, "De la abacá, que es la Musa textilis," 128–29; Madulid, "The Life and Work of Luis Née," 43–44.
70. Née, "De la abacá," 124–25.
71. Dacudao, *Abaca Frontier*, 224.
72. Spencer, "Abacá and the Philippines," 95.
73. Bureau of Insular Affairs, *Official Handbook, Description of the Philippines*, pt. 1, 119.
74. *A Pronouncing Gazetteer and Geographical Dictionary of the Philippine Islands, United States of America, with Maps, Charts, and Illustrations*, 71. One imperial ton is 2,240 pounds and differs from the US ton, which is 2,000. The metric ton, on the other hand, is 2,204 pounds.
75. Brown, "Philippine Fiber Plants," 365.
76. Brown, "Philippine Fiber Plants," 319; Spencer, 96.
77. Owen, *Prosperity without Progress*, 12–13.
78. Jose and Dacudao, "Visible Japanese and Invisible Filipino Narratives of the Development of Davao, 1900s to 1930s," 115–20.
79. Houston, "The Philippine Abaca Industry: 1934–1950," 412.
80. Owen, *Prosperity without Progress*, 80–81.
81. Quizon, "Between the Field and the Museum," 292.
82. Hayase, "Tribes on the Davao Frontier, 1899–1941," 140–47.
83. Dacudao, *Abaca Frontier*, 242.
84. "The Bagobo," lecture notes, pp. 18–19, box 1, photo lot 107, Elizabeth H. and Sarah S. Metcalf Collection, National Anthropological Archives, SI.
85. Smith, "Making and Knowing in a Sixteenth-Century Goldsmith's Workshop," 43–57.
86. J. Roberts, "On Mis-Expertise: The Art Historian in the Studio," paper, College Art Association Annual Conference, Los Angeles, February 2018, accessed July 26, 2022. https://www.academia.edu/41545307/On_Mis_Expertise_the_Art_Historian_in_the_Studio.
87. J. Roberts, "Things: Material Turn, Transnational Turn," 67. Roberts synthesizes Ingold's highly instructive "Toward an Ecology of Materials," 427–42.
88. Quizon, "Between the Field and the Museum," 294.

89. "Skirt, woman's panapisan," 70.1/5366, Laura Watson Benedict Collection, Division of Anthropology, AMNH.

90. Quizon, "Two Yankee Women at the St. Louis Fair," 547.

91. Sibayan, "The Laura Watson Benedict Bagobo Textile Collection," 336–39.

92. Aguilar, "Rice and Magic," 297–330.

93. Wernstedt and Simkins, "Migrations and the Settlement of Mindanao," 86–89.

94. Dacudao, *Abaca Frontier*, 168.

95. Wernstedt and Simkins, "Migrations and the Settlement of Mindanao," 86–89.

96. Benedict, "Tolus Ka Talegit," p. 2, Laura Watson Benedict Purchase, Archives of the Division of Anthropology, AMNH.

97. Photograph of the US Army Air Corps, box 2, photo lot 107, Elizabeth and Sarah S. Metcalf Collection, National Anthropological Archives, SI; Photograph of Bagobo colleagues of collector Sarah S. Metcalf, *Sunday Telegram*, September 18, 1932, box 1, photo lot 107, Elizabeth and Sarah S. Metcalf Collection, National Anthropological Archives, SI. For excellent coverage of photo-elicitation and more contemporary Bagobo response to the latter image, see Quizon, "Two Yankee Women," 548–49.

98. Benedict, *A Study of Bagobo Ceremonial Magic and Myth*, 5–6.

99. Dacudao, *Abaca Frontier*, 275–78. Dacudao cites Cole, *The Wild Tribes of Davao District*, 91. For a brief history of trading post development and practices on Navajo lands, see Bsumek, *Indian-Made*, 47–75.

100. Bright yellow cotton threads are woven into the tapang fabric of the central ine piece of a panapisan skirt. The cotton behaves accordingly with the abacá threads, standing out most prominently for their color. Panapsian skirt, E286120, Division of Anthropology, SI.

101. Benedict, "Tolus Ka Talegit," p. 2, Laura Watson Benedict Purchase, Archives of the Division of Anthropology, AMNH.

102. *Encyclopaedia Britannica*, "synthetic dyes," accessed June 7, 2022, https://www.britannica.com/technology/dye/Synthetic-dyes.

103. Brown, "Natural Dyes of the Philippines," 385.

104. Beaded skirts, E286154-0 and E286153-0, Division of Anthropology, SI.

105. Benedict, "Tolus Ka Talegit," p. 1, Laura Watson Benedict Purchase, Archives of the Division of Anthropology, AMNH.

106. Benedict, "The TITIG or LOCOB," ca. 1910s, 70.1, 1910-2_70.1/5250-7178, Laura Watson Benedict Purchase, Archives of the Division of Anthropology, AMNH.

107. Quizon, "Between the Field and the Museum," 294–95.

108. Sibayan, "The Laura Watson Benedict Collection," 336.

109. Benedict, *A Study of Bagobo Ceremonial, Magic, and Myth*, 210–40.

110. Dacudao, *Abaca Frontier*, 120–21.

111. Bernstein, "The Perils of Laura Watson Benedict," 173–74.

112. Bankoff, "Wants, Wages, and Workers," 83–86.

113. Dacudao, "Empire's Informal Ties," 194–95.

114. Hayase, "Tribes of the Davao Frontier," 148.

115. Dacudao, "Empire's Informal Ties," 192.

116. Bernstein, "The Perils of Laura Watson Benedict," 180–84.

117. Accession notes, Laura Watson Benedict Purchase, Archives of the Division of Anthropology, AMNH.

CHAPTER 5. FIELD LABOR'S MENACE

1. "Report of the Chief of the Bureau of Forestry of the Philippine Islands," 286.
2. Merrill, "Report of the Botanist," 710.
3. Schiebinger, *Plants and Empire*; Clarke, *Aboriginal Plant Collectors*; Musselman, "Plant Knowledge at the Cape."
4. Konishi, Nugent, and Shellam, "Exploration Archives and Indigenous Histories," 5.
5. Shapin, "The Invisible Technician."
6. Garforth, "In/Visibilities of Research: Seeing and Knowing in STS," 268–69.
7. Gómez-Barris, *The Extractive Zone*, 11–12.
8. Weston, *Specters of Germany*, 42–48.
9. Ultramar 526, exp. 3, núm. 1, AHN.
10. Konishi, Nugent, and Shellam, "Exploration Archives and Indigenous Histories," 7. The editors succinctly describe the argument found in Lübcke, "Encounters and the Photographic Record in British New Guinea," 5.
11. Quizon and Afable, "Guest Editors' Introduction," 439–44.
12. Rusert, *Fugitive Science*, 26. On the extractive view, see Gómez-Barris, *The Extractive Zone*, 5–9.
13. W. Anderson, "Filming Fore, Shooting Scientists," 114–15.
14. Shellam, "Mediating Encounters Through Bodies and Talk," 100. Shellam engages Miller, "History from Between," 610–13.
15. Bankoff, "Wants, Wages, and Workers," 61.
16. Merrill, "Report of the Botanist," 645.
17. Freer, "The Work of the Bureau of Government Laboratories, of the Philippine Islands," 105.
18. Worcester, "Report of the Secretary of the Interior to the Philippine Commission for the Year Ending August 31, 1902," 289–90.
19. N. Roberts, "U.S. Forestry in the Philippines," 81; Bankoff, "Breaking New Ground?," 378.
20. Ultramar, leg. 527, exp. 1, núm. 38, AHN.
21. Ultramar, leg. 528, exp. 6, núm. 3, AHN.
22. *Forestry Matters in the Philippines, Hearings Before Committee on Insular Affairs*, House of Representatives, 10.
23. "Laboratories of the Bureau of Science, Philippine Government," 144–54, AHC. The article does not have a byline, though it quotes Paul C. Freer at length.
24. W. Anderson, *Colonial Pathologies*, 7.
25. Stewart, "How Did They Die?," 49. Stewart notes that Jack died mysteriously at sea, though Merrill considered this claim "erroneous" and insisted that Jack perished from pulmonary tuberculosis while battling malaria. Succeeding biographies of Jack contain similar reports. See Merrill, "William Jack's Genera and Species of Malaysian Plants," 200.
26. Ultramar, leg. 524, exp. 9, núm. 10, AHN.
27. Stewart, "How Did They Die?," 49.
28. For a thorough study of "self-sacrifice" as it emerged in late nineteenth-century US scientific discourse, see Herzig, *Suffering for Science*.

29. Malkiel, *Strangers in Yemen*, 189; *Encyclopaedia Britannica*, 9th ed., under "Seetzen, Ulrich Jasper," 581. Accounts and memorials persist today to recognize naturalists who died in the field such as Conniff's "The Wall of the Dead."

30. Foreman, *The Philippine Islands*. Spelling original.

31. Stewart, "How Did They Die?," 49.

32. "Slain Americans Found," *New York Times*, July 12, 1908; Ingles, *1908: The Way It Really Was*, 178. Both quote from and refer to a telegram produced by Ahern.

33. Van Steenis et al., *Flora Malesiana* 1, 56.

34. N. Roberts, "U.S. Forestry in the Philippines," 322–23; see also E. D. Merrill to Director of Civil Service, November 29, 1932, folder 14—Philippines B-M, box 3, ser. 2, correspondence, Elmer D. Merrill Records, NYBG.

35. In an 1892 issue of *La Ilustración filipina*, illustrator Doroteo Vicente provided an image captioned "Mountain negritos who assisted the naturalist Vidal" to demonstrate different racial types of the colony. The image suggests that Vidal collaborated with Aeta (for the now rather defunct "Negrito" label), male and female. See Zaragoza, *La Ilustración filipina, 1891–1894*, 9.

36. Bankoff, "Wants, Wages, and Workers," 63–64.

37. Cerón, *Memoria suscrita por el Illmo. Sr. Don Salvador Cerón y leida en el acto de la inauguración de la estatua del Exmo. Sr. D. Sebastián Vidal y Sloler*, HL-UST.

38. Bankoff, "Wants, Wages, and Workers," 84–85.

39. "Isotype of Canarium Ahernianum Merr. [Family Burseraceae]," February 1904, in JSTOR Global Plants. JSTOR.

40. "Ramos, Maximo (–1932)," JSTOR Global Plants. A Spaniard and specialist of mollusks is also presumed to have collected under the synonym "Ahern's Collector," but the proportion of collections attributed to him is much smaller. See "Quadras, José Florencio (fl. 1890–1901)," JSTOR Global Plants; Van Steenis et al., *Flora Malesiania*, 420.

41. E. D. Merrill to Secretary of Agriculture and Natural Resources, April 26, 1920, ser. 2, box 3, folder 14, Merrill Papers, LTML-NYBG.

42. M. Clemens to E. D. Merrill, May 3, 1926, Clemens Correspondence, box 5, University Herbarium, BHL.

43. Pancho, introduction to *Vascular Flora of Mount Makiling and Vicinity*, 10.

44. Frank Caleb Gates to the President of the University of the Philippines, August 26, 1913, Frank C. Gates Papers, BHL.

45. N. Roberts, "U.S. Forestry in the Philippines," 340.

46. Fedor Jagor's 1873 *Reisen in den Philippinen* (Travels in the Philippines) featured an image of what Jagor described as a "Bicol Naturalist," who likely assisted the German in the field. Nathaniel Parker Weston intimates that Jagor's titling of the field assistant is more "tongue in cheek" than earnest. See Weston, *Specters of Germany*, 45–46.

47. Bankoff, "Wants, Wages, and Workers," 70.

48. Merrill to Secretary of Agriculture and Natural Resources, April 26, 1920, Merrill Papers, LTML-NYBG.

49. Merrill, "New Plants from Sorsogon Province, Luzon," 2.

50. Merrill to Secretary of Agriculture and Natural Resources, April 26, 1920, Merrill Papers, LTML-NYBG.

51. "Laboratories of the Bureau of Science, Philippine Government," 144–54.

52. Rafael, *Contracting Colonialism*, 111–15.

53. Retana, *Supersticiones de los indios filipinos: un libro de aniterías*. Other similar inventories include Villaverde, *Supersticiones y cuentos de los Igorrotes*; Tenorio et al., *Costumbres de los indios tirurayes*.

54. Reyes, "Science, Sex, and Superstition," 92–93.

55. Calderon, "Tropical Obstetrical Problems," 383.

56. Laktaw, *Diccionario Tag'alog-Hispano*, 563.

57. De los Reyes, *El Folk-lore filipino*, 65.

58. De los Reyes, *La religión antigua de los Filipinos*. See also de los Reyes, *La religión del Katipunan*, from which the chapter's epigraph also comes.

59. Thomas, *Orientalists, Propagandists, and Ilustrados*, 90–91.

60. Paterno, *Nínay (costumbres filipinas)*, 159–69. Aswang, as one Visayan allegedly recounted to a US observer, had been spiritual and political counsel to precolonial datu, defamed with malevolence and falsehood during Christianity's ascent across the islands. See Millington and Maxfield, "Philippine (Visayan) Superstitions," 207.

61. Hamilton, Subramaniam, and Willey, "What Indians and Indians Can Teach Us about Colonization," 615.

62. "Ang aral ng mga mason," *Patnubay nğ Bayan*, no. 19, p. 12.

63. "Ang mga taong bago," *Pagsulong sa Karunungan*, no. 8, p. 1.

64. "Ang pagsasaka," *Patnubay nğ Bayan*, no. 44, pp. 25–26.

65. "Hula o pamahiin ng mga Astrologo," *Patnubay nğ Bayan* 3, no. 23, pp. 12–13; "Hula o pamahiin ng mga Astrologo," *Patnubay nğ Bayan* 3, no. 24, p. 24.

66. Paglinawan, *Balarilang Tagalog*, 303.

67. Wheatley, "US Colonial Governance of Superstition and Fanaticism in the Philippines," 25–33.

68. Meñez, "Encounters with Spirits," 252–53.

69. Ghosh, *The Nutmeg's Curse*, 87. Ghosh references Ehrenreich, *Desert Notebooks*.

70. Gardner, "Philippine (Tagalog) Superstitions," 191.

71. Benedict, *A Study of Bagobo Ceremonial, Magic and Myth*, 33.

72. Gardner, "Philippine (Tagalog) Superstitions," 201.

73. US Bureau of Insular Affairs, *A Pronouncing Gazetteer and Geographical Dictionary of the Philippine Islands*, 680.

74. Worcester, *The Philippine Islands and Their People*, 268. Wheatley also refers to this example in "US Colonial Governance of Superstition," 26.

75. Merrill, "New Philippine Moraceae," 55.

76. *Report of the Forestry Bureau*, 481. I thank Maia Jardenil Mislang for discovering this source.

77. Gibbs, "The Alcohol Industry of the Philippine Islands, Part I," 173.

78. Mueggler, *The Paper Road*, 85–86.

79. For example, shamans and other indigenous informants created fake plant names and equivocated about plants' properties as Francisco Hernández (1514–1587) worked to compile a major natural history of central Viceroyalty of New Spain in the sixteenth century. See Cañizares-Esguerra, *How to Write the History of the New World*, 63–64. Contemporary botanists also speak of misdirections and fake names in the very recent past.

80. M. S. Clemens to Merrill, May 3, 1926, BHL.
81. Baker, "Exploring with Aborigines," 37.
82. Clemens to Merrill, May 3, 1926, BHL.
83. Schiebinger, *Plants and Empire*, 68.
84. "Chit-Chat Clementis," June 26, 1925, unaddressed correspondence, Mary Stuart [*sic*] Clemens—1920–1929, BHL.
85. E. Quisumbing to E. D. Merrill, April 26, 1930; Eduardo Quisumbing to E. D. Merrill, August 19, 1930; Quisumbing to E .D. Merrill, December 9, 1930; "Begoniaceae loaned to Dr. I. Irmscher," April 24, 1930, all in ser. 2, folder 14, box 3, Merrill Records.
86. E. Quisumbing to E. D. Merrill, May 24, 1932, ser. 2, folder 15, box 3, Merrill Records.
87. Quisumbing to Merrill. *Science* published an obituary that appeared in the "In Memoriam" section, attributed to an unnamed correspondent, 27–28.
88. Hermes G. Gutierrez, in-person communication with author, December 17, 2017.
89. Hartman, "Venus in Two Acts," 14.
90. Spalink, "Taphonomic Historiography," 86.
91. Martin, *Inventing Superstition*, 126.
92. Martin, *Inventing Superstition*, 127–30; Scheid, "superstitio," in *Oxford Classical Dictionary*, March 7, 2016, https://doi.org/10.1093/acrefore/9780199381135.013.6150.
93. Martin, *Inventing Superstition*, 1–2.

CHAPTER 6. THE LATIN BABBLE

1. M. Clemens to E. D. Merrill, May 3, 1926, Clemens Correspondence, 1920–1929, box 5, University Herbarium, BHL. (Hereafter Clemens Correspondence.)
2. M. Clemens to Merrill, February 25, 1928, Clemens Correspondence, BHL.
3. M. Clemens to Merrill, March 2, 1928, Clemens–E. D. Merrill 1925–1929, box 5, University Herbarium, BHL. Emphasis original.
4. Culley, "Why vouchers matter in botanical research."
5. Van Steenis et al., *Flora Malesiana*, 10–11.
6. Schiebinger, *Plants and Empire*, 194–95.
7. Gordin, *Scientific Babel*, 25.
8. Daston, "Type Specimens and Scientific Memory," 199–201.
9. Gordin, *Scientific Babel*, 33–48.
10. *Cambridge Advanced Learner's Dictionary*, under "babble."
11. Candolle, *Laws of Botanical Nomenclature*, 12.
12. Urban, "Entextualization, Replication, and Power," 21.
13. Madulid, *A Dictionary of Philippine Plant Names*, 169.
14. Merrill, "Autobiography," unpublished manuscript, n.p. 1930, 1–16, LTML-NYBG. (Hereafter "Autobiography.")
15. Merrill, "Autobiography," 16–17.
16. W. Anderson, "Science in the Philippines," 299. Anderson cites Stanley, *A Nation in the Making*, and Owen, *Compadre Colonialism*. See also Freer, "The Work of the Bureau of Government Laboratories," 105.
17. Merrill, "Autobiography," 23–24.

18. Merrill, "Autobiography," 19.
19. Merrill, *Botanical Work in the Philippines*, 5–22.
20. Merrill, *A Dictionary of the Native Plant Names of the Philippine Islands*, 7–9.
21. Lee, "Between Universalism and Regionalism," 661–84.
22. Bartlett, lecture manuscript, "Nationalism, Imperialism, and Spheres of Influence in Natural Science," 40, folder 3—concerning botanical subjects, box 7—correspondence, Harley H. Bartlett Papers, BHL.
23. Mickulas, *Britton's Botanical Empire*, 158.
24. Merrill, *An Interpretation of Rumphius's Herbarium Amboinense*, 43.
25. Merrill, *Species Blancoanae*, 35.
26. Ikenberry, "Rev. Joseph Clemens, 1862–1936"; Clemens, Joseph (1862–1936) MC2001.17, DC. MC 2001.17, Joseph Clemens Diaries, ASCDC.
27. J. Clemens, diary entry, April 6, 1902, Transcript, Diary 1902, MC2001.17, Joseph Clemens Diaries, ASCDC.
28. Welch, "Atrocities in the Philippines," 238–89.
29. J. Clemens, diary entry, April 12, 1902, Joseph Clemens Diaries.
30. M. Clemens to Howe, September 3, 1936, Mary Strong Clemens, 1930–1939, H. H. Bartlett Correspondence, box 5, University Herbarium, BHL.
31. Merrill, "New Philippine Plants from the Collections of Mary Strong Clemens, I," 129–30.
32. Ikenberry, "Rev. Joseph Clemens, 1862–1936," Joseph Clemens Diaries.
33. M. Clemens to Merrill, January 7, 1929, Clemens—E .D. Merrill—1925–1929, Clemens Correspondence.
34. Von Zinnenburg Carroll, *Botanical Drift*, 42–47.
35. Handwritten caption on photograph of Mary Strong Clemens, unnamed porter, and Joseph Clemens, I Yüan K'ou, Chihli Province, April 28, 1913, Photographs 1912–1913, box 1, Mary Strong Clemens Personal Papers, NYBG.
36. M. Clemens to Patterson, September 1, 1922, Clemens–Patterson (USDA), 1921–1922, box 6, University Herbarium, BHL.
37. Secord, "Corresponding Interests," 388. Secord engages Rudwick, *The Great Devonian Controversy*. See also N. Johnson, "On the Colonial Frontier," 427.
38. M. Clemens to Bartlett, August 12, 1935, H. H. Bartlett Correspondence, Clemens Correspondence.
39. M. Clemens to Merrill, April 4, 1924, H. H. Bartlett Correspondence, Clemens Correspondence.
40. M. Clemens to Merrill, May 9, 1927, Clemens Correspondence. While most correspondence from this period does not include Annamese/Vietnamese diacritics, I include diacritics as they appear in the correspondence and publications and apply them to proper nouns.
41. Merrill to M. Clemens, May 2, 1927, Clemens Correspondence.
42. Merrill to M. Clemens, June 13, 1927, Clemens Correspondence.
43. Mary S. Clemens to Elmer D. Merrill, May 9, 1927, BHL.
44. Merrill, "Loureiro and His Botanical Work," 229. In addition to *Flora de Filipinas*, Blanco's interest in Philippine medicinal plants is especially evident in his 1831 work, *Ang mahusay na paraan nang pag-gamot sa manga may saquit ayon sa aral ni Tissot* (Efficient

remedies to treat the sick according to Tissot's teachings) published in Sampaloc by Cayetano J. Enriquez.

45. Merrill, "A Commentary," 7.
46. Merrill, "1919 'Commentary' on Loureiro's *Flora cochinchinensis*," xii–xxxvii, manuscript, 1919, ser. 3, Elmer Drew Merrill Records, NYBG. (Hereafter "1919 'Commentary.'")
47. Schiebinger, *Plants and Empire*, 220–23; Zarucchi, "The Treatment of Aublet's Generic Names," 215–42.
48. Bil, "Imperial Vernacular," 635–37.
49. Colmeiro, *Enumeración y revisión de las plantas de la Península Hispano-Lusitana é islas Baleares*, viii.
50. Merrill, "1919 'Commentary,'" xxxv–xxxvi.
51. Candolle, *Laws of Botanical Nomenclature*, 12.
52. Merrill, "Nomenclatural Notes on Rafinesque Published Papers 1804–1840," 203.
53. Merrill, "1919 'Commentary,'" xxxv–xxxvi.
54. Planta, *Traditional Medicine in the Colonial Philippines*, 143–203.
55. Merrill, "A Commentary on Loureiro's *Flora cochinchinensis*," 130–41, manuscript, 1934, ser. 3, Elmer Drew Merrill Records, NYBG. (Hereafter "1934 'Commentary.'") I compare this entry against Merrill's published 1935 "Commentary," 43–45.
56. Merrill, "1934 'Commentary,'" 137.
57. Yuncker, "The Fifth International Botanical Congress," 65.
58. Mickulas, *Britton's Botanical Empire*, 208.
59. Merrill, "1935 'Commentary,'" 34.
60. Sirks to National Research Council, memorandum, August 20, 1934, Merrill Records.
61. Schiebinger, *Plants and Empire*, 224.
62. Sirks to National Research Council, August 1934.
63. Schiebinger, *Plants and Empire*, 224.
64. Pulle to Merrill, May 25, 1934, Netherlands—ser. 2, folder 9, correspondence, Merrill Records.
65. "Botanical Collaboration Elmer D. Merrill," Netherlands—ser. 2, folder 9, box 3, Merrill Records.
66. Noé, "Review of *Chronica Botanica: Vol. 1*, edited by Fr. Verdoorn," 430.
67. Verdoorn to Merrill, August 7, 1934, Netherlands—ser. 2, folder 9, correspondence, Merrill Records.
68. Verdoorn to Merrill, December 5, 1934, Netherlands—ser. 2, folder 9, correspondence, Merrill Records.
69. Went to Merrill, March 28, 1934, Netherlands—ser. 2, folder 9, correspondence, Merrill Records.
70. Merrill, "The Cost of German Scientific Journals," 316.
71. Matera and Kent, "Introduction: The Wilsonian Moment Betrayed, 1919–1929," 3–4. In this climate, Verdoorn has also been described as one of several "scientific internationalists" who emerged. See Theunissen, "Unifying Science and Human Culture," 193. Somsen also describes an incident at the time of the outbreak of World War I upon the release of *An die Kulturwelt!* (To the civilized world!); scientist-detractors claimed that it espoused

violent German nationalism. See Somsen, "A History of Universalism: Conceptions of the Internationality of Science from the Enlightenment to the Cold War," 366–67.

72. Gleason, "Thumbnail Sketches of Botanists" 1961, 57–58, unpublished memoir, NYBG.

73. Merrill, "Autobiography," 88, LTML-NYBG.

74. Gutierrez, "*Cycas wadei* and Enduring White Space."

75. Merrill, "1935 'Commentary,'" 45.

76. J. Clemens to Merrill, June 16, 1927, Clemens Correspondence.

77. J. Clemens to Merrill, June 16, 1927.

78. M. Clemens to Merrill, July 11, 1927.

79. J. Clemens to Merrill, June 16, 1927.

80. J. Clemens and M. Clemens to Merrill, June 25, 1927.

81. J. Clemens and M. Clemens to Merrill, June 29, 1927.

82. J. Clemens to Merrill, August 5, 1927.

83. Merrill, too, critiqued her collecting sets but usually communicated his displeasure directly. For a representative example see Sydow to Brown, April 2, 1927, H. H. Bartlett Correspondence, Clemens Correspondence.

84. Merrill to M. Clemens, November 25, 1934, Mt. Kinabalu Collections, correspondence and package labels, box 1, Mary Strong Clemens Papers, NYBG.

85. Keeney, *The Botanizers*, 147–49.

86. M. Clemens to Merrill, August 5, 1927, Clemens Correspondence.

87. M. Clemens to Merrill, August 3, 1927. These annotations do not join the existing digitized specimens from Clemens's Indochina expedition and if included, were likely on appended sheets.

88. *Wrightia annamensis* Eberh. & Dubard, Annam, Tourane, J. Clemens with M. S. Clemens. Muséum national d'Histoire naturelle, Paris, Collection: Vascular plants (P), Specimen P00371380. The collecting annotation reads, "Cf. Wrightia. Red fl'd [flowered] shrub, 4–6 ft. along new road by cabin." "Cf." or "compare" indicates that Clemens suggests to her collaborators that the specimen might belong to the *Wrightia* genus.

89. Merrill, "1935 'Commentary,'" 315.

90. Merrill, "1935 'Commentary,'" 225–361.

91. Raheja, "Caste, Colonialism, and the Speech of the Colonized," 494–95.

92. Lumbera, *Tagalog Poetry*, 2.

93. Bowker and Star, *Sorting Things Out*, 41.

94. Yun, "Negotiating a Korean National Myth," 311. Yun engages Kondo, *Crafting Selves*, 311.

95. Mackay, "Agents of Empire," 38–44.

96. J. Clemens, diary entry, April 12, 1902, MC2001.17, Joseph Clemens Diaries, ASCDC.

97. Kramer, *The Blood of Government*, 87–158.

98. Ikenberry, *Mary Strong Clemens, a Botanical Pilgrimage*.

99. M. Clemens to Landis, March 11, 1927, H. H. Bartlett Correspondence, Clemens Correspondence.

100. J. and M. Clemens to Merrill, June 29, 1927, Clemens Correspondence.

101. Tran, *Gods, Heroes, and Ancestors*.

102. Yun, "Negotiating a Korean National Myth," 295.
103. Gutierrez, "Rehabilitating Botany in the Postwar Moment," 61–64.

CONCLUSION. OF PLACE, MOMENT, AND SOURCE

1. "Ukol sa mga Bulaklak," *Renacimiento Filipino: Seccion Tagala* 29 (1911): 33. My translation.
2. Cañizares-Esguerra, *Nature, Empire, and Nation*, 23. Cañizares-Esguerra works from Richard L. Kagan's concept of *arcana imperii* found in Kagan et al., "*Arcana Imperii*," 49–70. Raquel A. G. Reyes also summarizes Kagan's argument in Reyes, "Collecting and the Pursuit of Scientific Accuracy," 77–78.
3. Freer, "The Philippine Journal of Science," 1.
4. For a significant extant account from García on nineteenth-century work completed under the Spanish, see "Brief Review of the Forestry Service during the Spanish Government. From 1863 to 1898. By Regino García, an official of the Philippine Botanical and Forest Service from 1866 to 1898. Entered Forestry Bureau May, 1900," Insular Bureau of Forestry, box 586, Gifford Pinchot Papers, Library of Congress.
5. Ultramar, leg. 527, exp. 1, núm. 73 [labeled 63], AHN. An exception existed for the arts, as discussed in chapter 2.
6. Cooper, *Inventing the Indigenous*, 116–40.
7. Cañizares-Esguerra, *Nature, Empire, and Nation*; see also Cañizares-Esguerra, "Screwing Humboldt."
8. Ebach, *Origins of Biogeography*, 125.
9. Van Steenis et al., *Flora Malesiana*, vii.
10. Van Steenis et al., *Flora Malesiana*, xxxiv.
11. Van Steenis et al., *Flora Malesiana*, 10–11.
12. Ebach, *Origins of Biogeography*, 16–17.
13. Cañizares-Esguerra, "Screwing Humboldt."
14. Medrano, *The Edible Ocean*.
15. Emmerson, "'Southeast Asia': What's in a Name?," 6–7. British lawyer and ethnologist James Richardson Logan (1819–69) initiated the scholarly use of "Southeast Asia" in 1847.
16. Mojares, *Brains of the Nation*, 31.
17. Ultramar, leg. 526, exp. 8, núm. 89, AHN. For secondary accounts of the 1897 fire that destroyed records of the Jardín Botánico de Manila, see also Bankoff, "A Month in the Life of José Salud," 9.
18. Merrill, *Botanical Work in the Philippines*, 21.
19. Bankoff, "A Month in the Life," 19; N. Roberts, "U.S. Forestry in the Philippines," 77.
20. Ahern, *Important Philippine Woods*, 13.
21. N. Roberts, "U.S. Forestry in the Philippines," 318.
22. Bankoff, "'The Tree as the Enemy of Man,'" 390.
23. Merrill, *Botanical Work in the Philippines*, 35.
24. N. Roberts, "U.S. Forestry in the Philippines," 318.
25. McCoy, Scarano, and Johnson, "On the Tropic of Cancer," 11–12.
26. McCoy, Scarano, and Johnson, "On the Tropic of Cancer," 12.

27. Chao, "In the Shadow of the Palm," 638. Chao usefully engages Star and Griesemer's "boundary object" from "Institutional Ecology, 'Translations,' and Boundary Objects." See also Chao, *In the Shadow of the Palms*.

28. Dacudao, *Abaca Frontier*, 355.

29. Gieryn, *Truth-Spots: How Places Make People Believe*.

30. Ultramar, leg. 527, exp. 2, núm. 7, AHN.

31. I credit the ongoing work of STS scholar Elaine Gan, whose writing attends to the "multispecies temporalities" of *Oryza sativa*.

32. Rafael, "Nationalism, Imagery and the Filipino Intelligentsia," in *Discrepant Histories*, 136.

33. Rafael, "Nationalism, Imagery and the Filipino Intelligentsia," 153–54.

34. Aquino, "Veto Message of President Aquino on House Bill No. 5655."

35. Cepeda, "House Pushes Bills Declaring Waling-Waling, Balangay as National Symbols"; Congress of the Philippines, Senate Bill No. 3307, An Act Declaring the Waling-Waling Orchids as National Flower of the Philippines in Addition to Sampaguita; Jazul, "PH Can Have a Second National Flower after Sampaguita—National Museum Chief."

36. My usage of "provincializing" comes from Chakrabarty, *Provincializing Europe*.

37. Debates are ongoing, but important readings include Gillman and Wright, "Restoring Indigenous Names in Taxonomy"; Smith, Figueiredo, Hammer, and Thiele, "Dealing with Inappropriate Honorifics in a Structured and Defensible Way Is Possible," 933–35; and Thiele, Smith, Figueiredo, and Hammer, "Taxonomists Have an Opportunity to Rid Botanical Nomenclature of Inappropriate Honorifics in a Structured and Defensible Way," 1151–54.

38. Mosyakin, "If 'Rhodes-' Must Fall, Who Shall Fall Next?," 249–55.

39. See also Smith et al., "Dealing with Inappropriate Honorifics," 934.

40. "El arbolado de paseo en Manila," *Revista de Filipinas* 2, no. 1 (1876): 7, HL-UST.

41. Exemplary studies include Kroupa, "*Ex epistulis Philippinensibus*," 229–59; and "Georg Joseph Kamel (1661–1706)" (PhD diss., Cambridge University, 2019); Planta, *Traditional Medicine in the Colonial Philippines*.

42. The work of the PANTROPOCENE group of the Max Planck Institute for Geoanthropology is especially compelling alongside the community-engaged archaeological work under Stephen Acabado, the genomics research on rice under Michael Purugganan's laboratory at New York University, and relatedly, the Philippine Rice Genome Project funded by the Philippine Department of Agriculture and the Department of Science and Technology.

43. Rafael, preface to the paperback edition, *Contracting Colonialism*, xiv.

44. Afable, "Journeys from Bontoc to the Western Fairs, 1904–1915," 448.

Bibliography

ARCHIVAL COLLECTIONS AND HERBARIA

I have included parenthetical abbreviations of select archives or repositories. These abbreviations appear in the chapter endnotes for ease of reference.

Bagobo Collections. Division of Anthropology, Smithsonian Institution, Washington, DC (SI).
Bartlett, Harley H. Collection. Bentley Historical Library, University of Michigan, Ann Arbor (BHL).
Benedict, Laura Watson. Collection. Division of Anthropology, American Museum of Natural History, New York, NY (AMNH).
Benedict, Laura Watson. Purchase collection. Archives of the Division of Anthropology, American Museum of Natural History, New York, NY (AMNH).
Biblioteca Digital Hispánica. Digital repository. Biblioteca Nacional de España, Madrid, Spain. https://bdh.bne.es.
Far Eastern Review. Periodicals. American Historical Collection, Rizal Library, Ateneo de Manila University, Quezon City, Metro Manila, Philippines (AHC).
Gates, Frank C. Papers. Bentley Historical Library, University of Michigan, Ann Arbor (BHL).
HathiTrust Digital Library. Ann Arbor, MI. https://www.hathitrust.org/the-collection.
International Botanical Congress. Vertical files. LuEsther T. Mertz Library, New York Botanical Garden, Bronx (LTML-NYBG).
JSTOR Global Plants. JSTOR digital database. https://plants.jstor.org.
Linnaean Herbarium. Department of Phanerogamic Botany, Swedish Museum of Natural History, Stockholm (S-LINN).
Lopez Memorial Museum. Image Collection. Pasig, Metro Manila, Philippines.
Manila Daily Bulletin. Periodicals. American Historical Collection, Rizal Library, Ateneo de Manila University, Quezon City, Metro Manila, Philippines (AHC).
Manila Times. Periodicals. American Historical Collection, Rizal Library, Ateneo de Manila University, Quezon City, Metro Manila, Philippines (AHC).

Mapas, Planos y Dibujos (MPD). Government Records. Archivo Histórico Nacional, Madrid, Spain (AHN).
Memorias. National Archives of the Philippines, Manila (NAP).
Merrill, Elmer Drew. Records. Archives of the New York Botanical Garden, Bronx (NYBG).
Metcalf, Elizabeth, and Sarah S. Collection. National Anthropological Archives, Smithsonian Institution, Washington, DC (SI).
Museo Nacional del Prado. Madrid, Spain.
Pardo de Tavera Room and Special Collections. Rizal Library, Ateneo de Manila University, Quezon City, Metro Manila, Philippines.
Pinchot, Gifford. Papers. Library of Congress, Washington, DC.
Renacimiento Filipino. Periodicals. Heritage Library, Miguel de Benavides Library, University of Santo Tomas, Manila, Philippines (HL-UST).
Retrato Collection. Filipinas Heritage Library. Ayala Museum, Makati, Metro Manila, Philippines.
Revista de Filipinas. Periodicals. Heritage Library, Miguel de Benavides Library, University of Santo Tomas, Manila, Philippines (HL-UST).
Rolfe, Robert Allen. Directors' Correspondence. Library and Archives at Royal Botanic Gardens, Kew, London, UK.
Sloane, Hans. Manuscripts. British Library, London, UK.
Ultramar, Legajos 524, 526, 527, 528, and 534. Government Records. Archivo Histórico Nacional, Madrid, Spain (AHN).
University Herbarium. Records. Bentley Historical Library, University of Michigan, Ann Arbor (BHL).
Varios Personajes. National Archives of the Philippines, Manila (NAP).
Vascular Plants Collection. Muséum national d'Histoire naturelle, Paris, France.
Vidal y Soler, Sebastián. Collection. Herbario, Real Jardín Botánico de Madrid, Spain.

PUBLISHED PRIMARY SOURCES

Adanson, Michel. *Familles des plantes*. Paris: Vincent, 1763.
Ahern, George Patrick. *Important Philippine Woods: Compilation of Notes on the Most Important Timber Tree Species of the Philippine Islands*. Manila: Bureau of Forestry, 1903.
Aiton, William. *Hortus Kewensis, or, A Catalogue of Plants Cultivated in the Royal Botanic Garden at Kew*. London: Georg Nicol, 1789.
Alcina, Francisco Ignacio. *History of the Bisayan People in the Philippine Islands: Evangelization and Culture at the Contact Period*. Translated by Cantius J. Kobak and Lucio Gutiérrez. Manila: University of Santo Tomas Publishing House, (1668) 2002.
"Ang aral ng mga Mason." *Patnubay nğ Bayan* 2, no. 19 (1915).
"Ang mga taong bago." *Pagsulong sa Karunungan* 4, no. 8 (1911).
"Ang pagsasaka." *Patnubay nğ Bayan* 4, no. 44 (1918).
Aquino III, Benigno S. "Veto Message of President Aquino on House Bill No. 5655." *Official Gazette of the Government of the Philippines*, March 26, 2013. https://www.officialgazette.gov.ph/2013/03/26/veto-message-of-president-aquino-on-house-bill-no-5655/.

Benedict, Laura Watson. *A Study of Bagobo Ceremonial, Magic and Myth*. Netherlands: E. J. Brill, 1916.

Benedict, Laura Watson. "A Study of Bagobo Ceremonial, Magic and Myth." *Annals of the New York Academy of Sciences* 25, no. 1 (1916): 1–308.

Blair, Emma, and James Robertson, eds. *The Philippine Islands*. Vol. 20. Cleveland: Arthur H. Clark, 1904.

Blanco, Francisco Manuel. *Flora de Filipinas según el sistema sexual de Linneo*. Manila: Candido Lopez, 1837.

Brown, William Henry. "Natural Dyes of the Philippines." In Brown, *Minor Products*.

Brown, William Henry. Philippine Fiber Plants. In Brown, *Minor Products*.

Brown, William Henry. Preface to *Minor Products*.

Brown, William Henry, ed. *Minor Products of Philippine Forests*. Vols. 1–2, *Bureau of Forestry Bulletin*, no. 22. Manila: Bureau of Printing, 1920.

Brown, William Henry, and A. F. Fischer. "Philippine Forest Products as Sources of Paper Pulp." In Brown, *Minor Products*.

Brown, William Henry, and E. D. Merrill. "Philippine Fiber Plants." In Brown, *Minor Products*.

Bureau of Forestry. *Report of the Forestry Bureau of the Philippine Islands*. Manila: Bureau of Insular Affairs, War Department, 1902.

Bureau of Insular Affairs. *Description of the Philippines: Official Handbook and Catalogue of the Philippine Exhibit*, pt. 1. Washington, DC: Bureau of Public Printing, 1903.

Bureau of Insular Affairs. *A Pronouncing Gazetteer and Geographical Dictionary of the Philippine Islands, United States of America, with Maps, Charts and Illustrations. Also the Law of Civil Government in the Philippine Islands Passed by Congress and Approved by the President July 1, 1902, with a Complete Index*. Washington, DC: Government Printing Office, 1902.

Calderon, Fernando. "Tropical Obstetrical Problems." *Philippine Journal of Science* 10, no. 6 (1915): 371–83.

Candolle, Alphonse de, and International Botanical Congress. *Laws of Botanical Nomenclature Adopted by the International Botanical Congress, Held at Paris in August 1867; Together with an Historical Introduction and a Commentary*. London: L. Reeve, 1868.

Census of the Philippine Islands, Taken under the Direction of the Philippine Commission in the Year 1903. Vol. 3. Washington, DC: US Bureau of the Census, 1905.

Cepeda, Mara. "House Pushes Bills Declaring Waling-Waling, Balangay as National Symbols." *Rappler*, November 19, 2019. https://www.rappler.com/nation/245304-house-bill-declaring-waling-waling-balangay-national-symbols.

Cerón, Salvador. *Memoria suscrita por el Illmo. Sr. Don Salvador Cerón y leida en el acto de la inauguración de la estatua del Exmo. Sr. D. Sebastian Vidal y Soler*. Manila: E. Bota, 1892.

Chavez, Ana Maria. "The Sampaguita." *The Philippine Republic*, 1925.

Chicago Tribune. "Murdered by Natives: First Lieutenant Bolton, U.S.A., and Benjamin Christian Slain on West Coast of Davao, Mindanao." June 15, 1906.

Cole, Fay-Cooper. *The Wild Tribes of Davao District, Mindanao*. Vol. 12. Chicago: Field Museum of Natural History, 1913.

Colmeiro, Miguel. *Enumeración de las plantas de la Peninsula Hispano-Lusitana é Islas Baleares, con la distribución geográfica de las especies, y sus nombres vulgares, tanto nacionales como provinciales.* Vol. 3. Madrid: Fuentenebro, 1885.

Congress of the Philippines, Senate. An Act Declaring the Waling-Waling Orchids as National Flower of the Philippines in Addition to Sampaguita as Declared in Proclamation No. 652 Dated 01 February 1934 by Governor-General Frank Murphy, 3307 § (2012). https://www.senate.gov.ph/lisdata/1423611973!.pdf.

Copeland, Edwin Bingham. *The Polypodiaceae of the Philippine Islands: New Species of Edible Philippine Fungi.* Manila: Bureau of Public Printing, 1905.

Darwin, Charles. *On the Origin of Species by Means of Natural Selection, or, The Preservation of Favoured Races in the Struggle for Life.* London: J. Murray, 1859.

De los Reyes, Isabelo. *El Folk-lore filipino.* Manila: Chofré y Compañía, 1889.

De los Reyes, Isabelo. *La religión antigua de los Filipinos.* Manila: Impr. de el Renacimiento, 1909.

De los Reyes, Isabelo. *La religión del Katipunan o sea la antigua de los Filipinos.* 2nd ed. Madrid: Tipolit de J. Corrales, 1900.

Diderot, Denis, and Jean Le Rond d'Alembert. *Encyclopédie de Diderot et d'Alembert: ou Dictionnaire raisonné des sciences, des arts et des métiers.* Paris: Chez Briasson, 1765.

Exposición de Filipinas: colección de artículos publicados en El Globo, diario ilustrado, político, científico y literario. Madrid: Tip. de El Globo, 1887.

Foreman, John. *The Philippine Islands: A Political, Geographical, Ethnographical, Social and Commercial History of the Philippine Archipelago, Embracing the Whole Period of Spanish Rule.* London: S. Low, Marston and Co., 1899.

Freer, Paul C. "The Philippine Journal of Science." *Philippine Journal of Science* 1, no. 1 (1906): 1–2.

Freer, Paul C. "The Work of the Bureau of Government Laboratories, of the Philippine Islands." *Science* 20, no. 499 (1904): 105–9.

Gaceta de Madrid. "Ministerio de Hacienda." 1872, 210, sec. Administración Central.

García López, Rafael. *Orígen é historia del Jardín botánico y de la Escuela de agricultura de Filipinas.* Spain: J. Iniesta, 1872.

García y Baza, Regino. Prologue to *Los arboles de goma, resinas y frutos oleosos de Filipinas.* Manila: Colegio de Santo Tomas, 1902.

Gardner, Fletcher. "Philippine (Tagalog) Superstitions." *Journal of American Folklore* 19, no. 74 (1906): 191–204.

Gibbs, David. *Advanced English Grammar and Composition.* New York: American Book Company, 1908.

Gibbs, H. D. "The Alcohol Industry of the Philippine Islands, Part I." *Philippine Journal of Science* 6, no. 3 (1911): 99–205.

Gray, Asa. "XXV.—Do Varieties Wear Out, or Tend to Wear Out?" *Journal of Natural History* 15, no. 87 (2009): 192–97.

Gray, Asa. *Letters of Asa Gray.* Vol. 1. Edited by Jane Loring Gray. Boston: Houghton, Mifflin, 1893.

Gray, Asa. *Scientific Papers of Asa Gray.* Vol. 1. Selected by Charles Sprague Sargent. Boston: Houghton, Mifflin, 1889.

Hanson, Otto O. "The Great Hemp District of Davao." *Manila Times*, February 1910, p. 108.

Hernández, Antonio Flórez, and Rafael de Piquer y Martín-Cortés. *Crónica de La Exposición de Filipinas: estudio crítico-descriptivo*. Madrid: Manuel Ginés Hernández, 1887.
"Hula o pamahiin ng mga Astrologo." *Patnubay ng̃ Bayan* 3, no. 23 (1917).
"Hula o pamahiin ng mga Astrologo." *Patnubay ng̃ Bayan* 3, no. 24 (1917).
Jaena, Graciano López. "Filipinas en La Exposición Universal de Barcelona." Barcelona: Busquets and Vidal, 1888.
Jagor, Fedor. *Viajes por Filipinas*. Translated by Sebastián Vidal y Soler. Madrid: Aribau y Ca. sucesores de Rivadeneyra, 1875.
Jagor, Fedor. *Reisen in den Philippinen*. Berlin: Weidmannsche Buchhandlung, 1873.
Jazul, Noreen. "PH Can Have a Second National Flower after Sampaguita—National Museum Chief." *Manila Bulletin*, July 23, 2020. https://mb.com.ph/2020/07/23/ph-can-have-a-second-national-flower-after-sampaguita-national-museum-chief/.
Jordana y Morera, Ramón. *Memoria sobre el comercio de maderas en Filipinas*. Madrid: R. Rojas, 1894.
Jordana y Morera, Ramón. *Memoria sobre la producción de los montes públicos de Filipinas en el año económico de 1872–1873, elevada al Excmo Señor Ministro de Ultramar*. Madrid: Manuel Minuesa, 1874.
Kaempfer, Engelbert. *Amoenitatum exoticarum politico-physico-medicarum fasciculi V*. N.p., 1712.
Kamel, Georg Joseph, and Jacobum Petiver. "Tractatus de Plantis Philippinensibus Scandentibus." *Philosophical Transactions of the Royal Society* 24 (1704): 1707–22.
Kapra, Mario. "Ukol sa mga bulaklak." *Renacimiento Filipino*, February 7, 1911, Sección Tagala.
Kurz, Wilhelm Sulpiz. *Forest Flora of British Burma*. Calcutta: Office of the Superintendent of Government Printing, 1877.
"Laboratories of the Bureau of Science, Philippine Government." *Far Eastern Review* 2, no. 6 (1905): 144–54.
Laktaw, Pedro Serrano. *Diccionario Hispano-Tagalog*. Manila: La Opinion, 1889.
Laktaw, Pedro Serrano. *Diccionario Tag'alog-Hispano*. Manila: Islas Filipinas, 1914.
Linnaeus, Carl. *Linnaeus' Philosophia botanica*. Translated by Stephen Freer. Oxford: Oxford University Press, 2003. First published 1751.
Linnaeus, Carl. *Species plantarum*. Stockholm: Laurentii Salvii, 1753.
Linnaeus, Carl. *A System of Vegetables (Systema vegetabilium)*. Translated by a Botanical Society in Lichfield. London: J. Jackson for Leigh and Sotheby, 1783.
Linnaeus, Carl. *Systema naturae*. Stockholm: Laurentii Salvii, 1758.
Lisboa, Márcos de. *Vocabulario de la lengua Bicol*. Philippines: Est. Tip. del Colegio de Santo Tomas, 1865.
Los Angeles Herald. "The National Floral Emblem." December 14, 1893.
Martinez, Felix. "¡Sampaguita!" Originally published 1893. Reprinted in *La Ilustración Filipina, 1891–1894*, edited by Ramon Zaragoza. Manila: RAMAZA, 1992.
Martyn, Thomas. *The Language of Botany: Being a Dictionary of the Terms Made Use of in That Science, Principally by Linneus*. London: Printed for B. and J. White, Fleet-Street, 1793.
Martyn, Thomas. "XV. Observations on the Language of Botany." *Transactions of the Linnean Society of London* 1 (1791): 147–54.

McKinley, William. "Report of a Commission Appointed to Investigate Affairs in the Philippine Islands." Congressional Address. Washington, DC: Government Printing Office, 1900.

Mentrida, Alonso de. *Diccionario de la lengua Bisaya Hiligueina y Haraya de la Isla de Panay*. Philippines: Imp. de D. Manuel y de D. Felis S. Dayot, 1841.

Merrill, Elmer D. *Botanical Work in the Philippines*. Manila: Bureau of Public Printing, 1903.

Merrill, Elmer D. "A Commentary on Loureiro's 'Flora Cochinchinensis.'" *Transactions of the American Philosophical Society* 24, no. 2 (1935): 1–445.

Merrill, Elmer D. "The Cost of German Scientific Journals." *Science* 81, no. 2100 (1935): 316.

Merrill, Elmer D. *A Dictionary of the Native Plant Names of the Philippine Islands*. Manila: Bureau of Public Printing, 1903.

Merrill, Elmer D. *A Discussion and Bibliography of Philippine Flowering Plants*. Manila: Bureau of Printing, 1926.

Merrill, Elmer D. *An Interpretation of Rumphius' Herbarium Amboinense*. Manila: Bureau of Science, 1917.

Merrill, Elmer D. "Loureiro and His Botanical Work." *Proceedings of the American Philosophical Society* 72, no. 4 (1933): 229–39.

Merrill, Elmer D. "New Philippine Moraceae." *Philippine Journal of Science* 18 (1921): 49–69.

Merrill, Elmer D. "New Philippine Plants from the collections of Mary Strong Clemens I." *Philippine Journal of Science* 3, no. 3, sec. C, Botany (1908): 129–65.

Merrill, Elmer D. "New Plants from Sorsogon Province, Luzon." *Philippine Journal of Science*, no. 9 (1916): 1.

Merrill, Elmer D. "Nomenclatural Notes on Rafinesque's Published Papers 1804–1840." *Journal of the Arnold Arboretum* 29, no. 2 (1948): 202–214.

Merrill, Elmer D. "The Present Status of Botanical Exploration of the Philippines." *Philippine Journal of Science* 10, no. 3 (1915).

Merrill, Elmer D. "Report of Elmer D. Merrill, Botanist." In *Report of the Governor General of the Philippine Islands Pt. 2*. Washington, DC: Government Printing Office, 1904.

Merrill, Elmer D. "Report of the Botanist." In *Report to the Philippine Commission Pt. 2*. Washington, DC: Bureau of Government Printing, 1903.

Merrill, Elmer D. *Species Blancoanae: A Critical Revision of the Philippine Species of Plants Described by Blanco and by Llanos*. Manila: Bureau of Printing, 1918.

Merrill, Elmer D. "William Jack's Genera and Species of Malaysian Plants." *Journal of the Arnold Arboretum* 33, no. 3 (1952): 199–251.

Millington, W. H., and Berton L. Maxfield. "Philippine (Visayan) Superstitions." *Journal of American Folklore* 19, no. 74 (1906): 205–11.

Molina, Antonio. *Ang Kundiman ng Himagsikan*. Manila: Bureau of Printing, 1940.

"Murdered by Natives: First Lieutenant Bolton, U.S.A., and Benjamin Christian Slain on West Coast of Davao, Mindanao." *Santa Barbara Morning Press*, June 15, 1906.

National Geographic. "Our State Flowers: The Floral Emblems Chosen by the Commonwealths." Vol. 31, no. 6 (June 1917): 481–517.

Née, Luis. "De la abacá, que es la Musa textilis." *Anales de Historia Natural* 4, no. 11 (1801): 123–29.

New York Times. "Slain Americans Found; Teachers Everett and Wakeley Were Drugged by Philippine Tribesmen." July 12, 1908.

Noceda, Juan José de, and Pedro de Sanlucar. *Vocabulario de la lengua tagala, trabajado por varios sugetos doctos y graves, y últimamente añadido, corregido y coordinado por el P. Juan de Noceda y el P. Pedro de Sanlucar, de la Compañia de Jesus.* Manila: Ramírez y Giraudier, 1860. First published 1832.

Paglinawan, Mamerto. *Balarilang Tagalog.* Manila: Limbagang Magiting ni Honorio Lopez, 1910.

Pardo de Figueroa, José Emilio. *Algunos escritos del teniente de navio J.E.P. de F. (Pascual Lúcas de la Encina); or denados y anotados por el Doctor Thebussem.* Madrid: M. Rivadeneyra, 1873.

Pardo de Tavera, Trinidad. *Medicinal Plants of the Philippines.* Translated by Jerome B. Thomas Jr. Philadelphia: P. Blakiston's Son and Co., 1901.

Pardo de Tavera, Trinidad. *Plantas medicinales de Filipinas.* Madrid: B. Rico, 1892.

Pardo de Tavera, Trinidad. *El sanscrito en la lengua tagalog.* Paris: Faculté de médecine, A. Davy, 1887.

Paterno, Pedro Alejandro. *Nínay (costumbres filipinas).* Madrid: Impr. de Fortanet, 1885.

Persoon, C. H. *Synopsis methodica fungorum.* Pts. 1–2. Gottingen: Apud Henricum Dieterich, 1801.

Philippines Bureau of Forestry. *Report of the Chief of the Bureau of Forestry of the Philippine Islands for the Period from 1904/05.* Manila: Bureau of Public Printing, 1905.

Presupuesto general de gastos e ingresos del Estado para el año 1853. Madrid: Ministerio de Hacienda, España, 1852.

"Recent Deaths." *Science* 76, no. 1958 (1932): 27–28.

Retana, W. E. *Supersticiones de los indios filipinos: un libro de aniterías.* Madrid: M. Minuesa de los Ríos, 1894.

Rheede, Hendrick van. *Hortus indicus malabaricus.* Vol. 6. Amsterdam: Johannis van Someren and Johannis van Dyck, 1686.

Rizal, José. "De Rizal a Barrantes—réplica de Rizal a la crítica de Barrantes contra el 'Noli.'" In *Epistolario rizalino, 1887–1890*, edited by Teodoro M. Kalaw, 2:301–2. Manila: Bureau of Printing, 1931.

Rizal, José. *Noli me tangere.* 2nd ed. Caracas: Biblioteca Ayacucho, 1982. First published 1887.

Rodriguez, Eulogio B. "The Legend of the Sampaguita: The Filipino National Flower." *N.Y.K. Line*, 1930.

Roxas, Felix. *The World of Felix Roxas: Anecdotes and Reminiscences of a Manila Newspaper Columnist, 1926–36.* Translated by Angel Estrada and Vincente del Carmen. Manila: Filipiniana Book Guild, 1970.

Rumphius, Georg Eberhard. *Herbarium Amboinense.* Vol. 5. Amsterdam: Apud Fransicum Changuion, Joannem Catuffe, and Hermannum Uytwerf, 1757.

Steenis, C. G. G. J. van, M. J. van Steenis-Kruseman, Indonesia Departemen Pertanian, Kebun Raya Indonesia, and Lembaga Ilmu Pengetahuan Indonesia. *Flora Malesiana.* Vol. 1, ser. 1. Jakarta: Noordhoff-Kolff, 1950.

Tenorio, José, Guillermo Bennásar, and American Philosophical Society. *Costumbres de los indios tirurayes.* Manila: Tip. Amigos del Pais, 1892.

United States 60th Congress. *Hearings before Committee on Insular Affairs, House of Representatives. Forestry Matters in the Philippines*, George Patrick Ahern. Washington, DC: Government Printing Office, 1908.

United States Philippine Commission. An Act Providing for the Establishment of Government Laboratories for the Philippine Islands. U.S.C. 156 (July 1, 1901).

Vidal y Soler, Sebastián. *Memoria sobre el ramo de montes en las Islas Filipinas*. Philippines: Aribau y Ca., 1874.

Vidal y Soler, Sebastián. *Sinopsis de familias y generos de plantas leñosas de Filipinas, introducción a la flora forestal del archipiélago filipino*. Manila: Chofré y Compañía, 1883.

Villaverde, Juan. *Supersticiones y cuentos de los Igorrotes*. Manila: Modern, 1897.

Vincente, Doroteo. "Negritos montanos que ayudaron al naturalista Vidal." Originally published 1893. Reprinted in *La Ilustración Filipina, 1891–1894*, edited by Ramon Zaragoza. Manila: RAMAZA, 1992.

Worcester, Dean C. "Report of the Secretary of the Interior to the Philippine Commission for the Year Ending August 31, 1902." In *Annual Reports of the War Department for the Fiscal Year Ended June 30, 1902*, pt. 1. Washington, DC: Government Printing Office, 1903.

Worcester, Dean C. *The Philippine Islands and Their People; a Record of Personal Observation and Experience, with a Short Summary of the More Important Facts in the History of the Archipelago*. London: Macmillan, 1909.

Yuncker, Truman G. "The Fifth International Botanical Congress." *Proceedings of the Indiana Academy of Science* 40 (1931): 61–66.

UNPUBLISHED SECONDARY SOURCES AND THESES

Kroupa, Šebestián. "Georg Joseph Kamel (1661–1706): A Jesuit Pharmacist at the Frontiers of Colonial Empires." PhD diss., Cambridge University, 2019.

Medrano, Anthony. *The Edible Ocean: Science, Industry, and the Rise of Urban Southeast Asia*. Yale University Press, forthcoming.

Parco-De Castro, Eloisa. "Fostering Social Transformation through Philippine Secondary Education." PhD diss., University of Santo Tomas, 2015.

Reyes, José G. "¿Quién fué Don Regino García y Baza? Notas biográficas sobre este ilustre botánico filipino." Unpublished manuscript, Manila, 1940, HL-UST.

Roberts, Nathaniel E. "U.S. Forestry in the Philippines: Environment, Nationhood, and Empire, 1900–1937." PhD diss., University of Washington, 2013.

PUBLISHED SECONDARY SOURCES

Abinales, Patricio N. "American Rule and the Formation of Filipino 'Colonial Nationalism.'" *Japanese Journal of Southeast Asian Studies* 39, no. 4 (2002): 604–21.

Abinales, Patricio N., and Donna J. Amoroso. *State and Society in the Philippines*. Lanham, MD: Rowman and Littlefield, 2017.

Afable, Patricia O. "Journeys from Bontoc to the Western Fairs, 1904–1915: The 'Nikimalika' and Their Interpreters." *Philippine Studies* 52, no. 4 (2004): 445–73.

Afable, Patricia O., Constance de Monbrison, and Corazon S. Alvina. *Philippines: An Archipelago of Exchange*. English ed. Arles, France: Actes sud, 2013.

Agoncillo, Teodoro A. *Introduction to Filipino History*. Manila: Radiant Star, 1974.

Agoncillo, Teodoro A. *Malolos: The Crisis of the Republic*. Quezon City: University of the Philippines, 1960.

Agoncillo, Teodoro A. *Prelude to 1896*. Professorial Chair Lecture Series 2. Quezon City: University of the Philippines Press, 1974.

Aguilar, Filomeno V. "Rice and Magic: A Cultural History from the Precolonial World to the Present." *Philippine Studies: Historical and Ethnographic Viewpoints* 61, no. 3 (2013): 297–330.

Aguilar, Filomeno V. "Romancing Tropicality: 'Ilustrado' Portraits of the Climate in the Late Nineteenth Century." *Philippine Studies: Historical and Ethnographic Viewpoints* 64, no. 3/4 (2016): 417–54.

Aguilar, Filomeno V. "Tracing Origins: 'Ilustrado' Nationalism and the Racial Science of Migration Waves." *Journal of Asian Studies* 64, no. 3 (2005): 605–37.

Aguilera-Manzano, José María. "Havana's Botanical Garden in the Construction of Cuban National Identity." Working paper, Max Weber Programme. European University Institute, Fiesole, 2007. https://cadmus.eui.eu/handle/1814/7422.

Alexander, Simone A. James. "M/othering the Nation: Women's Bodies as Nationalist Trope in Edwidge Danticat's 'Breath, Eyes, Memory.'" *African American Review* 44, no. 3 (2011): 373–90.

Alfred, Taiaiake. *Peace, Power, Righteousness: An Indigenous Manifesto*. 2nd ed. Don Mills, ON: Oxford University Press, 2009.

Alfred, Taiaiake. "Sovereignty." In Barker, *Sovereignty Matters*.

Anderson, Benedict. *Imagined Communities: Reflections on the Origin and Spread of Nationalism*. Rev. ed. London: Verso, 2006.

Anderson, Benedict. *Why Counting Counts: A Study of Forms of Consciousness and Problems of Language in Noli Me Tangere and El Filibusterismo*. Quezon City: Ateneo de Manila University Press, 2008.

Anderson, Elizabeth. "Feminist Epistemology: An Interpretation and a Defense." *Hypatia* 10, no. 3, Analytic Feminism (1995): 50–84.

Anderson, Warwick. *Colonial Pathologies: American Tropical Medicine, Race, and Hygiene in the Philippines*. Durham, NC: Duke University Press, 2006.

Anderson, Warwick. "Filming Fore, Shooting Scientists: Medical Research, Experimental Filmmaking, and Documentary Cinema." *Visual Anthropology* 32, no. 2 (2019): 109–27.

Anderson, Warwick. "Introduction: Postcolonial Technoscience." *Social Studies of Science* 32, no. 5/6 (2002): 643–58.

Anderson, Warwick. "Science in the Philippines." *Philippine Studies* 55, no. 3 (2007): 287–318.

Baker, Don. "Exploring with Aborigines: Thomas Mitchell and His Aboriginal Guides." *Aboriginal History* 22 (2020): 36–50.

Balce, Nerissa. *Body Parts of Empire: Visual Abjection, Filipino Images, and the American Archive*. Ann Arbor: University of Michigan Press, 2016.

Bañas Llanos, María Belén. *Ang Pagbubukid ng Kalikasang: Una historia natural de Filipinas, Juan de Cuéllar, 1739?–1801*. Barcelona: Serbal, 2000.

Bankoff, Greg. "Breaking New Ground? Gifford Pinchot and the Birth of 'Empire Forestry' in the Philippines, 1900–1905." *Environment and History* 15, no. 3 (2009): 369–93.

Bankoff, Greg. "A Month in the Life of José Salud, Forester in the Spanish Philippines, July 1882." *Global Environment* 2, no. 3 (2009): 8–47.

Bankoff, Greg. "The Science of Nature and the Nature of Science in the Spanish and American Philippines." In *Cultivating the Colonies: Colonial States and Their Environmental Legacies*, edited by Christina Folke Ax, Niels Brimnes, Niklas Thode Jensen, and Karen Oslund, 78–108. Athens: Ohio University Press, 2011.

Bankoff, Greg. "'The Tree as the Enemy of Man': Changing Attitudes to the Forests of the Philippines, 1565–1898." *Philippine Studies* 52, no. 3 (2004): 320–44.

Bankoff, Greg. "Wants, Wages, and Workers." *Pacific Historical Review* 74, no. 1 (2005): 59–86.

Barad, Karen. *Meeting the Universe Halfway: Quantum Physics and the Entanglement of Matter and Meaning*. Durham, NC: Duke University Press, 2007.

Barbour, Karen. "Embodied Ways of Knowing." *Waikato Journal of Education* 10 (2004): 227–38.

Barker, Joanne, ed. *Critically Sovereign: Indigenous Gender, Sexuality, and Feminist Studies*. Durham, NC: Duke University Press, 2017.

Barker, Joanne. "Sovereignty." In *Keywords for Gender and Sexuality Studies*, edited by Keywords Feminist Editorial Collective, Kyla Wazana Tompkins, Aren Aizura, Aimee Bahng, Karma R. Chávez, Mishuana Goeman, and Amber J. Musser, 211–15. New York: New York University Press, 2021.

Barker, Joanne, ed. *Sovereignty Matters: Locations of Contestation and Possibility in Indigenous Struggles for Self-Determination*. Lincoln: University of Nebraska Press, 2006.

Barnard, Timothy P. *Nature's Colony: Empire, Nation and Environment in the Singapore Botanic Gardens*. Singapore: National University of Singapore Press, 2017.

Barnes, Gina L. "An Introduction to Buddhist Archaeology." *World Archaeology* 27, no. 2 (1995): 165–82.

Barrera-Osorio, Antonio. *Experiencing Nature: The Spanish American Empire and the Early Scientific Revolution*. Austin: University of Texas Press, 2010.

Berlin, Brent, Dennis E. Breedlove, and Peter H. Raven. "Covert Categories and Folk Taxonomies." *American Anthropologist* 70, no. 2 (1968): 290–99.

Bernstein, Jay H. "The Perils of Laura Watson Benedict: A Forgotten Pioneer in Anthropology." *Philippine Quarterly of Culture and Society* 13, no. 3 (1985): 171–97.

Besky, Sarah, and Johnathan Padwe. "Placing Plants in Territory." *Environment and Society: Advances in Research* 7 (2016).

Bil, Geoff. "Imperial Vernacular: Phytonymy, Philology and Disciplinarity in the Indo-Pacific, 1800–1900." *British Journal for the History of Science* 51, no. 4 (2018): 635–58.

Blanco, John D. *Frontier Constitutions: Christianity and Colonial Empire in the Nineteenth-Century Philippines*. Berkeley: University of California Press, 2009.

Bleichmar, Daniela. *Visible Empire: Botanical Expeditions and Visual Culture in the Hispanic Enlightenment*. Chicago: University of Chicago Press, 2012.

Bleichmar, Daniela. "Visible Empire: Scientific Expeditions and Visual Culture in the Hispanic Enlightenment." *Postcolonial Studies* 12, no. 4 (2009).

Blok, Anders. "War of the Whales: Post-Sovereign Science and Agonistic Cosmopolitics in Japanese-Global Whaling Assemblages." *Science, Technology, and Human Values* 36, no. 1 (2011): 55–81.

Bonneuil, Christophe. "The Manufacture of Species: Kew Gardens, the Empire and the Standardisation of Taxonomic Practices in Late 19th Century Botany." In *Instruments, Travel, and Science: Itineraries of Precision from the Seventeenth to the Twentieth Century*, edited by Marie-Noelle Bourguet, Christian Licoppe, and Heinz Otto Sibum, 189–215. Routledge Studies in the History of Science, Technology, and Medicine. London: Routledge, 2002.

Borromeo-Buehler, Soledad. "The 'Inquilinos' of Cavite: A Social Class in Nineteenth-Century Philippines." *Journal of Southeast Asian Studies* 16, no. 1 (1985): 69–98.

Bowker, Geoffrey C., and Susan Leigh Star. *Sorting Things Out: Classification and Its Consequences*. Cambridge, MA: MIT Press, 2000.

Braverman, Irus. *Planted Flags: Trees, Land, and Law in Israel/Palestine*. New York: Cambridge University Press, 2009.

Brillantes, M. P. "La Flor de Manila." In *Cultural Center of the Philippines Encyclopedia of Philippine Art*. Vol. 6, Philippine Music. Manila: Cultural Center of the Philippines, 1994.

Brown, Kirby. "Sovereignty." *Western American Literature* 53, no. 1 (2018): 81–90.

Bsumek, Erika Marie. *Indian-Made: Navajo Culture in the Marketplace, 1868–1940*. CultureAmerica. Lawrence: University Press of Kansas, 2008.

Buhain, Dominador D. *A History of Publishing in the Philippines*. Quezon City: Rex Bookstore, 1998.

Burns, Lucy Mae San Pablo. "Which Way to the Philippines? United Stages of Empire." In *Puro Arte*, vol. 9. New York: New York University Press, 2012.

Burzynski, Joseph. "The Timber Trade and the Growth of Manila, 1864–1881." *Philippine Studies* 50, no. 2 (2002): 168–92.

Butzer, Karl W. "From Columbus to Acosta: Science, Geography, and the New World." *Annals of the Association of American Geographers* 82, no. 3 (1992): 543–65. https://sites.utexas.edu/butzer/files/2017/03/Butzer-1992-ColumbustoAcosta.pdf.

Caleya, Paloma Blanco Fernández de. "Sebastián Vidal y Soler." Madrid: Real Academia de la Historia, 2020.

Camacho, Leonarda. *100 Taon: 100 Filipina sa digmaan at sa kapayapaan*. Quezon City: SBA Printers, 2000.

Cañizares-Esguerra, Jorge. "How to Write the History of the New World: Histories, Epistemologies, and Identities in the Eighteenth-Century Atlantic World." In *How to Write the History of the New World*. Stanford: Stanford University Press, 2001.

Cañizares-Esguerra, Jorge. *Nature, Empire, and Nation: Explorations of the History of Science in the Iberian World*. Stanford: Stanford University Press, 2006.

Cañizares-Esguerra, Jorge. "Screwing Humboldt and His Hagiographers." *Medium*, 2019. https://jorgecanizaresesguerra.medium.com/screwing-in-two-positions-82c2cc5b09db.

Canonigo, Mar, ed. *Tinig ng Bayan*. Manila: Tambuli, 1972.

Capistrano-Baker, Florina H. "Whither Art History? Whither Art History in the Non-Western World: Exploring the Other('s) Art Histories." *Art Bulletin* 97, no. 3 (2015): 246–57.

Carroll, Khadija von Zinnenburg. *Botanical Drift: Protagonists of the Invasive Herbarium*. Cambridge, MA: MIT Press, 2017.

Casey, Edward S. "How to Get from Space to Place in a Fairly Short Stretch of Time: Phenomenological Prelegomena." In *Senses of Place: Toward a Renewed Understanding of the Place-World*, edited by Steven Feld and Keith H. Basso, 13–52. Santa Fe, NM: School of American Research Press, 1996.

Castro, Christi-Anne. *Musical Renderings of the Philippine Nation*. New York: Oxford University Press, 2011.

Chakrabarti, Pratik, and Michael Worboys. "Science and Imperialism since 1870." In *The Cambridge History of Science*, edited by David C. Lindberg, Ronald L. Numbers, and Roy Porter, 9–31. Cambridge: Cambridge University Press, 2003.

Chakrabarty, Dipesh. *Provincializing Europe: Postcolonial Thought and Historical Difference*. Princeton Studies in Culture/Power/History. Princeton, NJ: Princeton University Press, 2000.

Chambers, David Wade, and Richard Gillespie. "Locality in the History of Science: Colonial Science, Technoscience, and Indigenous Knowledge." *Osiris* 15 (2000): 221–40.

Chao, Sophie. "In the Shadow of the Palm: Dispersed Ontologies among Marind, West Papua." *Cultural Anthropology* 33, no. 4 (2018): 621–49.

Chao, Sophie. *In the Shadow of the Palms: More-Than-Human Becomings in West Papua*. Durham, NC: Duke University Press, 2022.

Charbonneau, Oliver. "'A New West in Mindanao': Settler Fantasies on the U.S. Imperial Fringe." *Journal of the Gilded Age and Progressive Era* 18, no. 3 (2019): 304–23.

Charbonneau, Oliver. *Civilizational Imperatives: Americans, Moros, and the Colonial World*. Ithaca, NY: Cornell University Press, 2020.

Chater, A. O., R. K. Brummitt, and Christiaan Hendrik Persoon. "Subspecies in the Works of Christiaan Hendrik Persoon." *TAXON* 15, no. 4 (1966): 143–49.

Choy, Timothy. *Ecologies of Comparison: An Ethnography of Endangerment in Hong Kong*. Durham, NC: Duke University Press, 2011.

Clark, John. "The Worlding of the Asian Modern." In *Contemporary Asian Art and Exhibitions*, edited by Michelle Antoinette and Caroline Turner. Canberra: Australian National University Press, 2011.

Clarke, Philip. *Aboriginal Plant Collectors: Botanists and Australian Aboriginal People in the Nineteenth Century*. Kenthurst, New South Wales: Rosenberg, 2008.

Claudio, Lisandro E. *Jose Rizal: Liberalism and the Paradox of Coloniality*. Global Political Thinkers, edited by Harmut Behr and Felix Rösch. Cham, Switzerland: Palgrave Macmillan, 2018.

Clausen, Robert T. "On the Use of the Terms 'Subspecies' and 'Variety.'" *Rhodora* 43, no. 509 (1941): 157–67.

Comaroff, Jean, and John Comaroff. "Naturing the Nation: Aliens, Apocalypse and the Postcolonial State." *Journal of Southern African Studies* 27, no. 3 (2001): 627–51.

Conniff, Richard. "The Wall of the Dead: A Memorial to Fallen Naturalists." *Strange Behaviors* (blog), January 14, 2011. https://strangebehaviors.wordpress.com/2011/01/14/the-wall-of-the-dead.

Coo, Stephanie Marie R. *Clothing the Colony: Nineteenth-Century Philippine Sartorial Culture, 1820–1896*. Quezon City: Ateneo de Manila University Press, 2019.

Cooper, Alix. *Inventing the Indigenous: Local Knowledge and Natural History in Early Modern Europe*. Cambridge: Cambridge University Press, 2007.

Culley, Theresa M. "Why Vouchers Matter in Botanical Research." *Applications in Plant Sciences* 1, no. 11 (2013): 1300076.

Dacudao, Patricia Irene. *Abaca Frontier: The Socioeconomic and Cultural Transformation of Davao, 1898–1941*. Quezon City: Ateneo de Manila University Press, 2023.

Dacudao, Patricia Irene. "Empire's Informal Ties: Pioneer Anthropologists in Davao, 1904–1916." *Philippine Studies: Historical and Ethnographic Viewpoints* 68, no. 2 (2020): 179–209.

Dallal, Ahmad. *Islam, Science, and the Challenge of History*. New Haven, CT: Yale University Press, 2010.

"Darwin Correspondence Project." Accessed August 31, 2023. https://www.darwinproject.ac.uk.

Das, Veena. *Life and Words: Violence and the Descent into the Ordinary*. Berkeley: University of California Press, 2007.

Daston, Lorraine. "Type Specimens and Scientific Memory." *Critical Inquiry* 31, no. 1 (2004): 153–82.

Daston, Lorraine, and Peter Galison. *Objectivity*. New York: Zone, 2007.

De Jesús, Edilberto C. *The Tobacco Monopoly in the Philippines: Bureaucratic Enterprise and Social Change, 1766–1880*. Quezon City: Ateneo de Manila University Press, 1980.

De La Calle, Sierra. *Félix Martínez y Lorenzo e La Ilustración Filipina*. Valladolid, Spain: Cuadernos del Museo Oriental-Valladolid, 2015.

De Lima Grecco, Gabriela, and Sven Schuster. "Decolonizing Global History? A Latin American Perspective." *Journal of World History* 31, no. 2 (2020): 425–46.

De Vos, Paula. "An Herbal El Dorado: The Quest for Botanical Wealth in the Spanish Empire." *Endeavour* 27, no. 3 (2004): 117–21.

De Vos, Paula. "Research, Development, and Empire: State Support of Science in the Later Spanish Empire." *Colonial Latin American Review* 15, no. 1 (2006): 55–79.

De Vos, Paula. "The Science of Spices: Empiricism and Economic Botany in the Early Spanish Empire." *Journal of World History* 17, no. 4 (2006): 399–427.

Dery, Luis Camara. *Awit kay inang bayan: ang larawan ng Pilipinas ayon sa mga tula't kundiman na kinatha noong panahon ng himagsikan*. Manila: De La Salle University Press, 2003.

Díaz Pascual, Concha. "Modelos para la escuela de pintura de Manila." *Cuaderno de Sofonisba* (blog), March 12, 2019. http://cuadernodesofonisba.blogspot.com/2019/03/la-eleccion-de-modelos-para-la-academia.html.

Diaz-Trechuelo, Maria Lourdes. "Eighteenth Century Philippine Economy: Agriculture." *Philippine Studies* 14, no. 1 (1966): 65–126.

Doeppers, Daniel F. *Feeding Manila in Peace and War, 1850–1945*. Madison: University of Wisconsin Press, 2016.

Drayton, Richard Harry. *Nature's Government: Science, Imperial Britain, and the "Improvement" of the World*. New Haven, CT: Yale University Press, 2000.

Dupré, John. "Natural Kinds and Biological Taxa." *Philosophical Review* 90, no. 1 (1981): 66–90.

Eamon, William. "'Nuestros males no son constitucionales sino circunstanciales': The Black Legend and the History of Early Modern Spanish Science." *Colorado Review of Hispanic Studies* 7 (2009): 13–30.

Ebach, Malte Christian. *Origins of Biogeography: The Role of Biological Classification in Early Plant and Animal Geography*. New York: Springer, 2015.

Ehrenreich, Ben. *Desert Notebooks: A Road Map for the End of Time*. New York: Counterpoint, 2020.

Elizalde, María Dolores. "Imperial Transition in the Philippines: The Making of a Colonial Discourse about Spanish Rule." In *Endless Empire*, edited by McCoy, Fradera, and Jacobson.

Emmerson, Donald K. "'Southeast Asia': What's in a Name?" *Journal of Southeast Asian Studies* 15, no. 1 (1984): 1–21.

Englund, Harri. "Ethnography after Globalism: Migration and Emplacement in Malawi." *American Ethnologist* 29, no. 2 (2002): 261–86.

Esposito, John L., ed. *The Oxford History of Islam*. Oxford: Oxford University Press, 2000.

Expeditions at the Field Museum. "Philippine Biodiversity," 2023. https://expeditions.fieldmuseum.org/island-mammals/philippine-biodiversity.

Fermin, Jose D. *1904 World's Fair: The Filipino Experience*. Honolulu: University of Hawaii Press, 2006.

Fernandez, Doreen. *Palayok: Philippine Food through Time, on Site, in the Pot*. Makati City: Bookmark, 2000.

Fernandez, Doreen. "The World of Miguel Ruiz." In *Reflections on Philippine Culture and Society: Festschrift in Honor of William Henry Scott*, edited by Jesus T. Peralta and William Henry Scott. Quezon City: Ateneo de Manila University Press, 2001.

Ferrer, Ada. *Insurgent Cuba: Race, Nation, and Revolution, 1868–1898*. Chapel Hill: University of North Carolina Press, 1999.

Flores, Patrick D. *Painting History: Revisions in Philippine Colonial Art*. Quezon City: University of the Philippines Press, 1998.

Flores, Patrick D., Michelle Antoinette, and Caroline Turner. "Polytropic Philippine: Intimating the World in Pieces." In *Contemporary Asian Art and Exhibitions*, 47–66. Connectivities and World-Making. Canberra: Australian National University Press, 2014.

Fox, Frederick. "Philippine Vocational Education: 1860–1898." *Philippine Studies* 24, no. 3 (1976): 261–87.

Fradera, Josep María. *Colonias para después de un imperio*. Barcelona: Ediciones Bellaterra, 2005.

Fradera, Josep María. "Reading Imperial Transitions." In McCoy and Scarano, *Colonial Crucible*, 34–62.

Galera Gómez, Andrés, ed. *El arca de Neé: plantas recolectadas por el botánico Luis Neé durante la Expedición Malaspina*. Madrid: Consejo Superior de Investigaciones Cientificas, 2016.

Garforth, Lisa. "In/Visibilities of Research: Seeing and Knowing in STS." *Science, Technology, and Human Values* 37, no. 2 (2012): 264–85.

Gascoigne, John. *Science in the Service of Empire: Joseph Banks, the British State and the Uses of Science in the Age of Revolution*. Cambridge: Cambridge University Press, (1998) 2010.

George, Alain. "Direct Sea Trade Between Early Islamic Iraq and Tang China: From the Exchange of Goods to the Transmission of Ideas." *Journal of the Royal Asiatic Society* 25, no. 4 (2015): 579–624.

Ghiselin, Michael T. "A Radical Solution to the Species Problem." *Systematic Zoology* 23, no. 4 (1974): 536–44.

Ghosh, Amitav. *The Nutmeg's Curse: Parables for a Planet in Crisis*. Chicago: University of Chicago Press, 2021.

Gieryn, Thomas F. *Truth-Spots: How Places Make People Believe*. Chicago: University of Chicago Press, 2018.

Gillman, Len Norman, and Shane Donald Wright. "Restoring Indigenous Names in Taxonomy." *Communications Biology* 3, no. 1 (October 23, 2020): 1–3.

Goeman, Mishuana, and Jennifer Denetdale, eds. "Native Feminisms: Legacies, Interventions, and Indigenous Sovereignties." *Wicazō Ša Review* 24, no. 2 (2009): 9–13.

Gómez-Barris, Macarena. *The Extractive Zone: Social Ecologies and Decolonial Perspectives*. Dissident Acts. Durham, NC: Duke University Press, 2017.

Gordin, Michael D. *Scientific Babel: How Science Was Done Before and After Global English*. Chicago: University of Chicago Press, 2015.

Gowing, Peter. "Mandate in Moroland: The American Government of Muslim Filipinos, 1899–1920." PhD diss., Syracuse University, 1968.

Green, Gillian. "Angkor Vogue: Sculpted Evidence of Imported Luxury Textiles in the Courts of Kings and Temples." *Journal of the Economic and Social History of the Orient* 50, no. 4 (2007): 424–51.

Grindstaff, Beverly K. "Creating Identity: Exhibiting the Philippines at the 1904 Louisiana Purchase Exposition." *National Identities* 1, no. 3 (1999): 245–63.

Gutaker, Rafal M., Simon C. Groen, Emily S. Bellis, Jae Y. Choi, Inês S. Pires, R. Kyle Bocinsky, Emma R. Slayton, et al. "Genomic History and Ecology of the Geographic Spread of Rice." *Nature Plants* 6, no. 5 (2020): 492–502.

Gutierrez, Kathleen C. "Botanical Knowledge within Itneg Weaving and Dyeing: Tracking Contemporary Negotiations with Plant-Based Technologies." In *Anthropological, Mathematical Symmetry and Technical Characterisation of Cordillera Textiles Project*, edited by Analyn Salvador-Amores. Quezon City: University of the Philippines Press, 2019.

Gutierrez, Kathleen C. "*Cycas wadei* and Enduring White Space." In *Empire and Environment: Ecological Ruin in the Transpacific*, edited by Jeffrey Santa Ana, Heidi Amin-Hong, Rina Garcia Chua, and Xiaojing Zhou. Ann Arbor: University of Michigan Press, 2022.

Gutierrez, Kathleen C. "Emina María Jackson y Zaragoza (1858–?): Illustration of *Diospyros embryopteris* in the Third Edition of Manuel Blanco's *Flora de Filipinas* (*Flora of the Philippines*) (1877–1883)." In *Women in the History of Science*, edited by Hannah Wills, Sadie Harrison, Erika Jones, Farrah Lawrence-Mackey, and Rebecca Martin, 242–46. London: UCL Press, 2023.

Gutierrez, Kathleen C. "From Objects of Study to Worldmaking Beings: The History of Botany at the Corner of the Plant Turn." *History Compass* 21, no. 8 (2023): e12782.

Gutierrez, Kathleen C. "Rehabilitating Botany in the Postwar Moment: National Promise and the Encyclopedism of Eduardo Quisumbing's Medicinal Plants of the Philippines (1951)." *Asian Review of World Histories* 6, no. 1 (2018): 33–67.

Hamilton, Jennifer A., Banu Subramaniam, and Angela Willey. "What Indians and Indians Can Teach Us about Colonization: Feminist Science and Technology Studies, Epistemological Imperialism, and the Politics of Difference." *Feminist Studies* 43, no. 3 (2017): 612–24.

Haraway, Donna. "Situated Knowledges: The Science Question in Feminism and the Privilege of Partial Perspective." *Feminist Studies* 14, no. 3 (1988): 575–99.

Hartman, Saidiya. "Venus in Two Acts." *Small Axe* 12, no. 2 (2008): 1–14.

Hau, Caroline. *Necessary Fictions: Philippine Literature and the Nation, 1946–1980*. Quezon City: Ateneo University Press, 2000.

Hayase, Shinzo. "Tribes on the Davao Frontier, 1899–1941." *Philippine Studies* 33, no. 2 (1985): 139–50.

Herzig, Rebecca M. *Suffering for Science: Reason and Sacrifice in Modern America*. New Brunswick, NJ: Rutgers University Press, 2005.

Ho, Peng Yoke, and Frederick Peter Lisowksi. *A Brief History of Chinese Medicine*. 2nd ed. Singapore: World Scientific, 1997.

Hoeg, Jerry. "The Reception of Charles Darwin in Spain and the Problem of Abulia in Pío Baroja's Camino de Perfección." In *Modernity and Epistemology in Nineteenth-Century Spain: Fringe Discourses*, edited by Ryan A. Davis and Alicia Cerezo Paredes, Lanham, MD: Lexington, 2016.

Houston, Charles O. "The Philippine Abaca Industry: 1934–1950." *University of Manila Journal of East Asiatic Studies* 3 (1954): 267–86; 408–15.

Ikenberry, Nelda B. *Mary Strong Clemens: A Botanical Pilgrimage, Her Glorious Mission from Here to the Outback Via Southeast Asia, a Biography*. Fort Worth, TX: BRIT, Fort Worth Botanical Garden, 2021.

Ingles, Raul Rafael. *1908: The Way It Really Was: Historical Journal for the UP Centennial, 1908–2008*. Quezon City: University of the Philippines Press, 2008.

Ingold, Tim. "Toward an Ecology of Materials." *Annual Review of Anthropology* 41 (2012): 427–42.

Joaquin, Nick. *A Question of Heroes*. Mandaluyong, Philippines: Anvil, 2005. First published 1977.

Joaquin, Nick. Introduction to *Juan Luna, the Filipino as Painter*, by Albano Pilar Santiago. Manila: Eugenio Lopez Foundation, 1980.

Johnson, Courtney. "Alliance Imperialism and Anglo-American Power after 1898: The Origins of Open-Door Internationalism." In *Endless Empire*, edited by McCoy, Fradera, and Jacobson, 122–35.

Johnson, Nuala C. "On the Colonial Frontier: Gender, Exploration and Plant-Hunting on Mount Victoria in Early 20th-Century Burma." *Transactions of the Institute of British Geographers* 42, no. 3 (2017): 417–31.

Jose, Lydia N. Yu, and Patricia Irene Dacudao. "Visible Japanese and Invisible Filipino Narratives of the Development of Davao, 1900s to 1930s." *Philippine Studies* 63, no. 1 (2015): 101–29.

Junco, José Álvarez. "The Formation of Spanish Identity and Its Adaptation to the Age of Nations." *History and Memory* 14, no. 1–2: Images of a Contested Past (2002): 13–36.

Junker, Laura L. *Raiding, Trading, and Feasting: The Political Economy of Philippine Chiefdoms*. Honolulu: University of Hawaii Press, 1999.

Jurilla, Patricia May B. *Tagalog Bestsellers of the Twentieth Century: A History of the Book in the Philippines*. Quezon City: Ateneo de Manila University Press, 2009.

Kagan, Richard L. "Prescott's Paradigm: American Historical Scholarship and the Decline of Spain." *American Historical Review* 101, no. 2 (1996): 423–46.

Kagan, Richard L., Pereda Marías, and Fernando Marías. "Arcana imperii: mapas, ciencia y poder en la corte de Felipe IV." In *El atlas del Rey Planeta: la "Descripción de España y las costas y puertos de sus reinos" de Pedro Texeira (1634)*, 49–70. Hondarribia, Spain: Nerea, 2002.

Keeney, Elizabeth. *The Botanizers: Amateur Scientists in Nineteenth-Century America*. Chapel Hill: University of North Carolina Press, 1992.

Kelley, Theresa M. *Clandestine Marriage: Botany and Romantic Culture*. Baltimore: Johns Hopkins University Press, 2012.

Kimaid, Michael. Review of *The Nutmeg's Curse: Parables for a Planet in Crisis*, by Amitav Ghosh. *Historical Geography* 49, no. 1 (2021): 100–102.

King, Amy M. *Bloom: The Botanical Vernacular in the English Novel*. New York: Oxford University Press, 2003.

Klock, John S. "Agricultural and Forest Policies of the American Colonial Regime in Ifugao Territory, Luzon, Philippines, 1901–1945." *Philippine Quarterly of Culture and Society* 23, no. 1 (1995): 3–19.

Kondo, Dorinne K. *Crafting Selves: Power, Gender, and Discourses of Identity in a Japanese Workplace*. Chicago: University of Chicago Press, 1990.

Konishi, Shino, Maria Nugent, and Tiffany Shellam. "Exploration Archives and Indigenous Histories: An Introduction." In *Indigenous Intermediaries*, 1–10.

Konishi, Shino, Maria Nugent, and Tiffany Shellam, eds. *Indigenous Intermediaries: New Perspectives on Exploration Archives*. Aboriginal History Monographs. Canberra: Australian National University Press, 2015.

Kramer, Paul A. *The Blood of Government: Race, Empire, the United States, and the Philippines*. Chapel Hill: University of North Carolina Press, 2006.

Kroupa, Šebestián. "*Ex epistulis Philippinensibus*: Georg Joseph Kamel SJ (1661–1706) and His Correspondence Network." *Centaurus* 57, no. 4 (2015): 229–59.

Lee, Jung. "Between Universalism and Regionalism: Universal Systematics from Imperial Japan." *British Journal for the History of Science* 48, no. 4 (2015): 661–84.

Legarda, Benito Justo. *After the Galleons: Foreign Trade, Economic Change & Entrepreneurship in the Nineteenth Century Philippines*. Madison: University of Wisconsin Press, 1999.

Legarda, Benito Justo. "The Economic Background of Rizal's Time." *Philippine Review of Economics* 48, no. 2 (2012): 1–22.

León, Felipe Padilla de. "Poetry, Music and Social Consciousness." *Philippine Studies* 17, no. 2 (1969).

Lesch, John E., T Frängsmyr, J. L. Heilbron, and R. E. Rider. "Systematics and the Geometrical Spirit." In *The Quantifying Spirit of the Eighteenth Century*, 73–112. Oakland: University of California Press, 2020.

Lewis, Daniel. *The Feathery Tribe: Robert Ridgway and the Modern Study of Birds*. New Haven, CT: Yale University Press, 2012.

Lim, T. K. *Edible Medicinal and Non-Medicinal Plants*. Vol. 8, *Flowers*. Dordrecht, Netherlands: Springer Science and Business, 2014.

Llorca, Jaume Josa. "La historia natural en la España del siglo XIX: botánica y zoología." *Ayer*, no. 7 (1992): 109–52.

Lübcke, Antje. "Encounters and the Photographic Record in British New Guinea." In *Indigenous Intermediaries*, edited by Konishi, Nugent, and Shellam, 169–88.

Lumbera, Bienvenido. *Tagalog Poetry, 1570–1898: Tradition and Influences in Its Development*. Quezon City: Ateneo de Manila University Press, 1986.

Lumbera, Bienvenido. *Writing the Nation / Pag-akda ng bansa*. Quezon City: University of the Philippines, 2000.

Luyt, Brendan. "Empire Forestry and Its Failure in the Philippines: 1901–1941." *Journal of Southeast Asian Studies* 47, no. 1 (2016): 66–87.

Mackay, David. "Agents of Empire: The Banksian Collectors and the Evaluation of New Lands." In *Visions of Empire: Voyages, Botany, and Representations of Empire*, edited by David Philip Miller and Peter Hanns Reill, 38–48. New York: Cambridge University Press, 2010.

MacLeod, R. "On Visiting the 'Moving Metropolis': Reflections on the Architecture of Imperial Science." *Historical Records of Australian Science* 5, no. 3 (1980): 1–16.

Madulid, Domingo A. *A Dictionary of Philippine Plant Names*. Makati City: Bookmark, 2001.

Madulid, Domingo A. "The Life and Work of Luis Née, Botanist of the Malaspina Expedition." *Archives of Natural History* 16, no. 1 (1989): 33–48.

Mahābhārata Book Twelve. Vol. 3. *Peace Part 2: The Book of Liberation*. Translated by Alex Wynne. Clay Sanskrit Library 58. New York: New York University Press, 2009.

Malkiel, David. *Strangers in Yemen: Travel and Cultural Encounter among Jews, Christians and Muslims in the Colonial Era*. Boston: De Gruyter Oldenbourg, 2020.

Mandelkern, India. "Taste-Based Medicine." *Gastronomica* 15, no. 1 (2015): 8–21.

Maravall, José Antonio. *Culture of the Baroque: Analysis of a Historical Structure*. Theory and History of Literature 25. Minneapolis: University of Minnesota Press, 1986.

Martin, Dale B. *Inventing Superstition: From the Hipporatics to the Christians*. Cambridge, MA: Harvard University Press, 2004.

Mastnak, Tomaz, Julia Elyachar, and Tom Boellstorff. "Botanical Decolonization: Rethinking Native Plants." *Environment and Planning D: Society and Space* 32 (2013): 363–80.

Matera, Marc, and Susan Kingsley Kent. "Introduction: The Wilsonian Moment Betrayed, 1919–1929." In *The Global 1930s: The International Decade*, 1–14. London: Taylor and Francis, 2017.

Mazower, Mark. "The Strange Triumph of Human Rights, 1933–1950." *Historical Journal* 47, no. 2 (2004): 379–98.

McCormick, Thomas. "From Old Empire to New: The Changing Dynamics and Tactics of American Empire." In McCoy and Scarano, *Colonial Crucible*, 64–79.

McCoy, Alfred W. "Fatal Florescence: Europe's Decolonization and America's Decline." In *Endless Empire*, 3–40.

McCoy, Alfred W., and Francisco A. Scarano, eds. *Colonial Crucible: Empire in the Making of the Modern American State*. Madison: University of Wisconsin Press, 2009. https://hdl.handle.net/2027/heb.08751.

McCoy, Alfred W., Francisco A. Scarano, and Courtney Johnson. "On the Tropic of Cancer: Transitions and Transformations in the U.S. Imperial State." In *Colonial Crucible*, 3–33.

McCoy, Alfred W., Josep María Fradera, and Stephen Jacobson, eds. *Endless Empire: Spain's Retreat, Europe's Eclipse, America's Decline*. Madison: University of Wisconsin Press, 2012.

Mears, Leon A. *Rice Economy of the Philippines*. Quezon City: University of the Philippines Press, 1974.

Meñez, Herminia Q. "Encounters with Spirits: Mythology and the Ingkanto Syndrome in the Philippines." *Western Folklore* 37, no. 4 (1978): 249–65.

Mickulas, Peter. *Britton's Botanical Empire: The New York Botanical Garden and American Botany, 1888–1929*. New York: New York Botanical Garden Press, 2007.

Miller, David Philip. "History from Between." *Technology and Culture* 52, no. 3 (2011): 610–13.

Mishler, Brent D. "Species Are Not Uniquely Real Biological Entities." In *Contemporary Debates in Philosophy of Biology*, edited by Francisco J. Ayala and Robert Arp. Hoboken, NJ: Wiley, 2009.

Mitchell, Robert. "Cryptogamia." *European Romantic Review* 21, no. 5 (2010): 631–51.

Mojares, Resil B. *Brains of the Nation: Pedro Paterno, T. H. Pardo de Tavera, Isabelo de Los Reyes and the Production of Modern Knowledge*. Quezon City: Ateneo de Manila University Press, 2006.

Mojares, Resil B. "The Formation of Filipino Nationality under U.S. Colonial Rule." *Philippine Quarterly of Culture and Society* 36 (2006): 11–32.

Mojares, Resil B. *Isabelo's Archive*. Mandaluyong: Anvil, 2013.

Mojares, Resil B. "Jose Rizal in the World of German Anthropology." *Philippine Quarterly of Culture and Society* 41, no. 3–4 (2013): 163–94.

Monnais-Rousselot, Laurence, C. Michele Thompson, and Ayo Wahlberg. Introduction to *Southern Medicine for Southern People: Vietnamese Medicine in the Making*. Newcastle: Cambridge Scholars, 2012.

Moore, Randy. "Linnaeus and the Sex Lives of Plants." *American Biology Teacher* 59, no. 3 (1997): 132.

Mortimer, Loren Michael. "Kateri's Bones: Recovering an Indigenous Political Ecology of Healing along Kaniatarowanenneh, 1660–1701." *Native American and Indigenous Studies* 7, no. 2 (2020): 55–86.

Mosyakin, Sergei L. "If 'Rhodes-' Must Fall, Who Shall Fall Next?" *TAXON* 71, no. 2 (2022): 249–55.

Mueggler, Erik. *The Paper Road: Archive and Experience in the Botanical Exploration of West China and Tibet*. Berkeley: University of California Press, 2011.

Müller-Wille, Staffan. "Linnaeus and the Love Lives of Plants." In *Reproduction: Antiquity to the Present Day*, edited by N. Hopwood, R. Fleming, and L. Kassell, 305–18. Cambridge: Cambridge University Press, 2018.

Musselman, Elizabeth Green. "Plant Knowledge at the Cape: A Study in African and European Collaboration." *International Journal of African Historical Studies* 36, no. 2 (2003): 367–92.

Myers, Norman, Russell A. Mittermeier, Cristina G. Mittermeier, Gustavo A. B. da Fonseca, and Jennifer Kent. "Biodiversity Hotspots for Conservation Priorities." *Nature* 403, no. 6772 (2000): 853–58.

Nabhan, Gary Paul, and Sara St. Antoine. "The Loss of Floral and Faunal Story: The Extinction of Experience." In *The Biophilia Hypothesis*, edited by Stephen R. Kellert and Edward O. Wilson, 229–250. Washington, DC: Island / Shearwater, 1993.

Nadasdy, Paul. *Sovereignty's Entailments: First Nation State Formation in the Yukon*. Toronto: University of Toronto Press, 2017.

Nanda, Meera. "The Epistemic Charity of the Social Constructivist Critics of Science and Why the Third World Should Refuse the Offer." In *A House Built on Sand: Exposing Postmodernist Myths about Science*, edited by Noretta Koertge, 286–312. New York: Oxford University Press, 1998.

Netzorg, Morton J. "Books for Children in the Philippines: The Late Spanish Period." *Philippine Quarterly of Culture and Society* 10, no. 4 (1982): 282–99.

Nicolson, Dan H. "A History of Botanical Nomenclature." *Annals of the Missouri Botanical Garden* 78, no. 1 (1991): 33–56.

Noé, A. C. Review of *Chronica Botanica*, vol. 1, by Fr. Verdoorn. *Botanical Gazette* 97, no. 2 (1935): 430–31.

Noyes, Dorothy. "Humble Theory." *Journal of Folklore Research* 45, no. 1 (2008): 37–43.

Ogilvie, Brian W. *The Science of Describing: Natural History in Renaissance Europe*. Chicago: University of Chicago Press, 2008.

Orillaneda, Bobby C. "Maritime Trade in the Philippines During the 15th Century CE." *Moussons. Recherche en sciences humaines Sur l'Asie du Sud-Est*, no. 27 (2016): 83–100.

Orillos-Juan, Ma Florina Yamsuan. "Ang proyektong Chico River Hydroelectric Dam: Hamon ng kaunlaran at reaksyong bayan 1965–1986." *Malay* 21, no. 2 (2009).

Osborne, Michael A. "Acclimatizing the World: A History of the Paradigmatic Colonial Science." *Osiris* 15 (2000): 135–51.

Otálora-Luna, Fernando, and Elis Aldana. "The Beauty of Sensory Ecology." *History and Philosophy of the Life Sciences* 39, no. 3 (2017): 1–7.

Overfield, Richard A. *Science with Practice: Charles E. Bessey and the Maturing of American Botany*. Ames: Iowa State University Press, 1993.

Owen, Norman G., ed. *Compadre Colonialism: Studies in the Philippines under American Rule*. Ann Arbor: University of Michigan Press, 1971.

Owen, Norman G. *Prosperity Without Progress: Manila Hemp and Material Life in the Colonial Philippines*. Quezon City: Ateneo de Manila University Press, 1984.

Pancho, Juan V. Introduction to *Vascular Flora of Mount Makiling and Vicinity (Luzon, Philippines): Part 1*. Supplement 1. "Kalikasan, the Philippine Journal of Biology." Quezon City: New Mercury, 1983.

Panganiban, José Villa. *Diksyunaryo-Tesauro Pilipino-Ingles*. Manila: Manlapaz, 1972.

Paredes, Alyssa. "Experimental Science for the 'Bananapocalypse': Counter Politics in the Plantationocene." *Ethnos* 88, no. 4 (2023): 837–63.

Parreñas, Juno Salazar. "6. From Decolonial Indigenous Knowledges to Vernacular Ideas in Southeast Asia." *History and Theory* 59, no. 3 (2020): 413–20.

Pelser, Pieter B., Julie F. Barcelona, and Daniel L. Nickrent. Co's Digital Flora of the Philippines website: "About This Website." Updated January 31, 2013. https://www.philippineplants.org/General/AboutLeonard.html.

Pérez, Joaquín Fernández, and Alberto Gomis Blanco. "La Ceres española y la Ceres europea, dos proyectos agrobotánicos de Mariano La Gasca y Simón de Rojas Clemente." *Llull* 13, no. 25 (1990): 379–401.

Philip, Kavita. *Civilizing Natures: Race, Resources and Modernity in Colonial South India*. New Brunswick, NJ: Rutgers University Press, 2003.

"Philippine Biodiversity." *Expeditions at the Field Museum*. Accessed July 19, 2024. https://expeditions.fieldmuseum.org/island-mammals/philippine-biodiversity.

Phillips, Christopher J. "The Taste Machine: Sense, Subjectivity, and Statistics in the California Wine World." *Social Studies of Science* 46, no. 3 (2016): 461–81.

Picq, Manuela Lavinas. *Vernacular Sovereignties: Indigenous Women Challenging World Politics*. Tucson: University of Arizona Press, 2018.

Pilapil, Vicente R. "The Cause of the Philippine Revolution." *Pacific Historical Review* 34, no. 3 (1965): 249–64.

Planta, Maria Mercedes G. *Traditional Medicine in the Colonial Philippines: Sixteenth through Nineteenth Century*. Manila: University of the Philippines, 2017.

Quirino, Carlos. "Damian Domingo, Filipino Painter." *Philippine Studies* 9, no. 1 (1961): 78–96.

Quirino, Carlos. "Manila's School of Painting." *Philippine Studies* 15, no. 2 (1967): 348–53.

Quizon, Cherubim A. "Between the Field and the Museum: The Benedict Collection of Bagobo Abaca Ikat Textiles." *Textile Society of America Symposium Proceedings*, 1998.

Quizon, Cherubim A. "Two Yankee Women at the St. Louis Fair." *Philippine Studies* 52, no. 4 (2004): 527–55.

Quizon, Cherubim A., and Patricia O. Afable. "Guest Editors' Introduction: Rethinking Displays of Filipinos at St. Louis: Embracing Heartbreak and Irony." *Philippine Studies* 52, no. 4 (2004): 439–44.

Qureshi, Sadiah. "Science, Empire and Globalization in the Nineteenth Century." In *The Routledge Research Companion to Nineteenth-Century British Literature and Science*, edited by John Holmes and Sharon Ruston, 19–29. London: Taylor and Francis, 2017.

Raes, Niels, and Peter C. van Welzen. "The Demarcation and Internal Division of Flora Malesiana: 1857–Present." *Blumea: Biodiversity, Evolution and Biogeography of Plants* 54, no. 1/3 (2009): 6–10.

Rafael, Vicente L. *Contracting Colonialism: Translation and Christian Conversion in Tagalog Society under Early Spanish Rule*. Ithaca, NY: Cornell University Press, 1988.

Rafael, Vicente L. *Discrepant Histories: Translocal Essays on Filipino Cultures*. Philadelphia, PA: Temple University Press, 1995.

Rafael, Vicente L. "Nationalism, Imagery, and the Filipino Intelligentsia in the Nineteenth Century." *Critical Inquiry* 16, no. 3 (1990): 591–611.

Rafael, Vicente L. "The War of Translation: Colonial Education, American English, and Tagalog Slang in the Philippines." *Journal of Asian Studies* 74, no. 2 (2015).

Rafael, Vicente L. *White Love and Other Events in Filipino History*. Durham, NC: Duke University Press, 2014.

Raheja, Gloria Goodwin. "Caste, Colonialism, and the Speech of the Colonized: Entextualization and Disciplinary Control in India." *American Ethnologist* 23, no. 3 (1996): 494–513.

Ray, Sugata. *Climate Change and the Art of Devotion: Geoaesthetics in the Land of Krishna, 1550–1850*. Seattle: University of Washington Press, 2019.

Reedy, David, Will C. McClatchey, Clifford Smith, Y. Han Lau, and K. W. Bridges. "A Mouthful of Diversity: Knowledge of Cider Apple Cultivars in the United Kingdom and Northwest United States." *Economic Botany* 63, no. 1 (2009): 2–15.

Reid, Anthony. "Violence at Sea: Unpacking 'Piracy' in the Claims of States over Asian Seas." In *Elusive Pirates, Pervasive Smugglers*, edited by Robert J. Antony. Hong Kong: Hong Kong University Press, 2010.

Restrepo, Eduardo, and Axel Rojas. *Inflexión decolonial: fuentes, conceptos y cuestionamientos*. Popayán, Colombia: Instituto de Estudios Sociales y Culturales Pensar, Maestría en Estudios Culturales, Universidad Javeriana; Editorial Universidad del Cauca, 2010.

Reyes, Raquel A. G. "Collecting and the Pursuit of Scientific Accuracy: The Malaspina Expedition in the Philippines, 1792." In *Empire and Science in the Making: Dutch Colonial Scholarship in Comparative Global Perspective, 1760–1830*, edited by P. Boomgaard. New York: Palgrave Macmillan, 2013.

Reyes, Raquel A. G. "Glimpsing Southeast Asian *Naturalia* in Global Trade, c. 300 BCE–1600 CE." In *Environment, Trade and Society in Southeast Asia: A Longue Durée Perspective*, edited by David Henley and Henk Schulte Nordholt, 96–119. Boston: Brill, 2015. https://doi.org/10.1163/9789004288058_008.

Reyes, Raquel A. G. *Love, Passion and Patriotism: Sexuality and the Philippine Propaganda Movement, 1882–1892*. Singapore: National University of Singapore Press, 2008.

Reyes, Raquel A. G. "Science, Sex and Superstition: Midwifery in 19th Century Philippines." In *Global Movements, Local Concerns: Medicine and Health in Southeast Asia*, edited by Laurence Monnais and Harold J. Cook, 81–103. Singapore: National University of Singapore Press, 2013.

Rice, Mark. *Dean Worcester's Fantasy Islands: Photography, Film, and the Colonial Philippines*. Ann Arbor: University of Michigan Press, 2014.

Robbins, Jane. "A Nation Within? Indigenous Peoples, Representation and Sovereignty in Australia." *Ethnicities* 10, no. 2 (2010): 257–74.

Roberts, Jennifer L. "On Mis-Expertise: The Art Historian in the Studio." Paper presented at the College Art Association Annual Conference, Los Angeles, 2018. Accessed July 19, 2024. https://www.academia.edu/41545307/On_Mis_Expertise_the_Art_Historian_in_the_Studio.

Roberts, Jennifer L. "Things: Material Turn, Transnational Turn." *American Art* 31, no. 2 (2017): 64–69.

Roces, Marian Pastor. "Costumes, colleciónes del trajes, costumbres, costumbrista." *Mapping Philippine Material Culture*. Accessed July 19, 2024. https://philippinestudies.uk/mapping/tours/show/21.

Roces, Marian Pastor, Dick Baldovino, and Wig Tysmans. *Sinaunang Habi: Philippine Ancestral Weave*. The Nikki Coseteng Filipiniana Series. Quezon City: N. Coseteng, 1991.

Rodogno, Davide. *Against Massacre: Humanitarian Interventions in the Ottoman Empire, 1815–1914*. Human Rights and Crimes against Humanity. Princeton, NJ: Princeton University Press, 2011.

Rodriguez, Lorenzo. "El Jardín Botánico de Manila y Don Sebastián Vidal y Soler." *Anales de La Real Academia de Farmacia* 28, no. 1–2 (1962).

Rodríguez, Rebeca Fernández. "Lexicography in the Philippines (1600–1800)." *Historiographia Linguistica* 41, no. 1 (2014): 1–32.

Rudwick, Martin J. S. *The Great Devonian Controversy: The Shaping of Scientific Knowledge among Gentlemanly Specialists*. Chicago: University of Chicago Press, 1985.

Rusert, Britt. *Fugitive Science: Empiricism and Freedom in Early African American Culture*. New York: New York University Press, 2017.

Rutherford, Danilyn. *Laughing at Leviathan: Sovereignty and Audience in West Papua*. Chicago: University of Chicago Press, 2012.

Ryan, John C. "Banyan." In *The Mind of Plants: Narratives of Vegetal Intelligence*, edited by John C. Ryan, Patrícia Vieria, Monica Gagliano, and Dennis McKenna, 21–29. Santa Fe, NM: Synergetic, 2021.

Rydell, Robert W. *All the World's a Fair: Visions of Empire at American International Expositions, 1876–1916*. Reprint edition. Chicago: University of Chicago Press, 1987.

Salvador-Amores, Analyn, ed. *Anthropological, Mathematical Symmetry and Technical Characterisation of Cordillera Textiles Project*. Quezon City: University of the Philippines Press, 2019.

Sánchez, Luis Angel Gómez. "La etnografía de filipinas desde la administración colonial española (1874–1898)." *Revista de Indias* 47, no. 179 (1987): 157–85.

Sánchez, Luis Angel Gómez. *Un imperio en la vitrina: el colonialismo español en el Pacífico y la Exposición de Filipinas de 1887*. Philippines: Consejo Superior de Investigaciones Científicas, Instituto de Historia, Departamento de Historia de América, 2003.

Sánchez, Luis Angel Gómez. "Indigenous Art at the Philippine Exposition of 1887: Arguments for an Ideological and Racial Battle in a Colonial Context." *Journal of the History of Collections* 14, no. 2 (2002): 283–94.

Santiago, Luciano P. R. "Damian Domingo and the First Philippine Art Academy (1821–1834)." *Philippine Quarterly of Culture and Society* 19, no. 4 (1991): 264–80.

Santiago, Luciano P. R. "The Flowering Pen: Filipino Women Writers and Publishers during the Spanish Period, 1590–1898, a Preliminary Survey." In "The Book," special issue, *Philippine Studies* 51, no. 4 (2003): 558–98.

Santiago, Luciano P. R. "The Painters of *Flora de Filipinas* (1877–1883)." *Philippine Quarterly of Culture and Society* 21, no. 2 (1993): 87–112.

Santiago, Luciano P. R., "Philippine Academic Art: The Second Phase (1845–98)." *Philippine Quarterly of Culture and Society* 17, no. 1 (1989): 67–89.

Santos, Boaventura de Sousa. *Epistemologies of the South: Justice against Epistemicide*. New York: Routledge, 2014.

Santos, Boaventura de Sousa, João Arriscado Nunes, and Maria Paula Meneses. "Opening Up the Canon of Knowledge and Recognition of Difference." In *Another Knowledge*

Is Possible: Beyond Northern Epistemologies, edited by Boaventura de Sousa Santos. London: Verso, 2007.

Schaeffer, Felicity Amaya. *Unsettled Borders: The Militarized Surveillance on Sacred Indigenous Land*. Durham, NC: Duke University Press, 2022.

Schiebinger, Londa L. *Plants and Empire: Colonial Bioprospecting in the Atlantic World*. Cambridge, MA: Harvard University Press, 2007.

Schmidt-Nowara, Christopher. *The Conquest of History: Spanish Colonialism and National Histories in the Nineteenth Century*. Pittsburgh: University of Pittsburgh Press, 2006.

Schumacher, John. "The Manila Synodal Tradition: A Brief History." *Philippine Studies* 27, no. 3 (1979): 285–348.

Schwartzberg, Joseph E. "Cosmography in Southeast Asia." In *The History of Cartography*, vol. 2, edited by J. B. Harley and David Woodward. Chicago: University of Chicago Press, 1994.

Scott, William Henry. "Class Structure in the Unhispanized Philippines." Special issue, *Philippine Studies* 27, no. 2 (1979): 139–59.

Scott, William Henry. *History on the Cordillera: Collected Writings on Mountain Province History*. Baguio: Baguio Printing and Publishing, 1975.

Scott, William Henry. "Sixteenth-Century Visayan Food and Farming." *Philippine Quarterly of Culture and Society* 18, no. 4 (1990): 291–311.

Secord, Anne. "Corresponding Interests: Artisans and Gentlemen in Nineteenth-Century Natural History." *British Journal for the History of Science* 27, no. 4 (1994): 383–408.

"Seetzen, Ulrich Jasper." In *Encyclopædia Britannica*. Vol. 24. Britannica Academic, 1911. http://archive.org/details/Encyclopaediabri24chisrich_201303.

Shapin, Steven. "The Invisible Technician." *American Scientist* 77, no. 6 (1989): 554–63.

Shapin, Steven. "Pump and Circumstance: Robert Boyle's Literary Technology." *Social Studies of Science* 14, no. 4 (1984).

Shellam, Tiffany. "Mediating Encounters through Bodies and Talk." In *Indigenous Intermediaries*, edited by Konishi, Nugent, and Shellam, 85–102.

Shimamura, Takanori. "What Is Vernacular Studies?" *Kwansei Gakuin University School of Sociology Journal* 129 (2018): 1–10.

Shteir, Ann B. *Cultivating Women, Cultivating Science: Flora's Daughters and Botany in England, 1760–1860*. Baltimore: Johns Hopkins University Press, 1996.

Sibayan, Judith Freya. "The Laura Watson Benedict Collection: Ancestral Clothing and the Agency of Things." In *Philippines: An Archipelago of Exchange*, edited by Constance de Monbrison and Corazon S. Alvina [English edition]. Arles: Actes sud, 2013.

Sigrist, René. "On Some Social Characteristics of the Eighteenth-Century Botanists." In *Scholars in Action: The Practice of Knowledge and the Figure of the Savant in the 18th Century*, edited by André Holenstein, Hubert Steinke, and Martin Stuber, vol. 1, 205–34. Boston: Brill, 2013.

Smith, Gideon F., Estrela Figueiredo, Timothy A. Hammer, and Kevin R. Thiele. "Dealing with Inappropriate Honorifics in a Structured and Defensible Way Is Possible." *TAXON* 71, no. 5 (2022): 933–35.

Smith, Pamela. "Making and Knowing in a Sixteenth-Century Goldsmith's Workshop." In *The Mindful Hand: Inquiry and Invention from the Late Renaissance to Early*

Industrialisation, edited by Lissa L. Roberts, Simon Schaffer, and Peter Dear, 43–57. Amsterdam: Royal Netherlands Academy of Arts and Sciences, 2007.

Smocovitis, Vassiliki Betty. "One Hundred Years of American Botany: A Short History of the Botanical Society of America." *American Journal of Botany* 93, no. 7 (2006): 942–52.

Somsen, Geert J. "A History of Universalism: Conceptions of the Internationally of Science from the Enlightenment to the Cold War." *Minerva* 46, no. 3 (2008): 361–79.

Spalink, Angenette. "Taphonomic Historiography: Excavating and Exhuming the Past in Suzan-Lori Parks's The America Play." *Modern Drama* 60, no. 1 (2017): 69–88.

Spencer, J. E. "Abacá and the Philippines." *Economic Geography* 27, no. 2 (1951): 95–106.

Spivak, Gayatri Chakravorty. "Can the Subaltern Speak?" In *Marxism and the Interpretation of Culture*, edited by Cary Nelson and Lawrence Grossberg. Urbana-Champaign: University of Illinois Press, 1988.

Stafleu, Frans Antoine. "Adanson and the Familles des plantes." In *Adanson: The Bicentennial of Michel Adanson's "Familles des plantes,"* edited by G. H. M. Lawrence, 123–264. Pittsburgh: Hunt Botanical Library, Carnegie Institute of Technology, 1963.

Stanley, Peter W. *A Nation in the Making: The Philippines and the United States, 1899–1921*. Harvard Studies in American-East Asian Relations 4. Cambridge, MA: Harvard University Press, 1974.

Star, Susan Leigh, and James R. Griesemer. "Institutional Ecology, 'Translations,' and Boundary Objects: Amateurs and Professionals in Berkeley's Museum of Vertebrate Zoology, 1907–39." *Social Studies of Science* 19, no. 3 (1989): 387–420.

Stewart, Ralph R. "How Did They Die?" *TAXON* 33, no. 1 (1984): 48–52.

Stoler, Ann Laura, and Frederick Cooper. "Between Metropole and Colony: Rethinking a Research Agenda." In *Tensions of Empire: Colonial Cultures in a Bourgeois World*, edited by Frederick Cooper and Ann Laura Stoler, 1–56. Oakland: University of California Press, 2019.

Subramaniam, Banu. *Ghost Stories for Darwin: The Science of Variation and the Politics of Diversity*. Urbana-Champaign: University of Illinois Press, 2014.

Subramaniam, Banu. "Science and Postcolonialism." In *A Companion to the History of American Science*, edited by Georgina M. Montgomery and Mark A. Largent, 491–501. Chichester, UK: Wiley-Blackwell, 2015.

Subramaniam, Banu, and Madelaine Bartlett. "Re-imagining Reproduction: The Queer Possibilities of Plants." *Integrative and Comparative Biology* 63, no. 4 (2023): 946–59. https://doi.org/10.1093/icb/icad012.

Tagliacozzo, Eric. *Secret Trades, Porous Borders: Smuggling and States along a Southeast Asian Frontier, 1865–1915*. New Haven, CT: Yale University Press, 2005.

Taiz, Lincoln, and Lee Taiz. *Flora Unveiled: The Discovery and Denial of Sex in Plants*. Oxford: Oxford University Press, 2017.

Taylor, Archer. "Characteristics of German Folklore Studies." *Journal of American Folklore* 74, no. 294 (1961): 293–301.

Taylor, Katherine Selena, Bradley J. Moggridge, and Anne Poelina. "Australian Indigenous Water Policy and the Impacts of the Ever-Changing Political Cycle." *Australian Journal of Water Resources* 20, no. 2 (2016): 132–47.

Thapar-Björket, Suruchi. "Gender, Nations and Nationalism." In *The Oxford Handbook of Gender and Politics*, edited by Georgina Waylen, 803–27. New York: Oxford University Press, 2013.

Theunissen, Bert. "Unifying Science and Human Culture: The Promotion of the History of Science by George Sarton and Frans Verdoorn." In *Pursuing the Unity of Science: Ideology and Scientific Practice from the Great War to the Cold War*, edited by Harmke Kamminga and Geert Somsen, 182–206. London: Routledge, 2016.

Thiele, Kevin R., Gideon F. Smith, Estrela Figueiredo, and Timothy A. Hammer. "Taxonomists Have an Opportunity to Rid Botanical Nomenclature of Inappropriate Honorifics in a Structured and Defensible Way." *TAXON* 71, no. 6 (2022): 1151–54.

Thomas, Deborah A., and Joseph Masco. *Sovereignty Unhinged: An Illustrated Primer for the Study of Present Intensities, Disavowals, and Temporal Derangements*. Durham, NC: Duke University Press, 2023.

Thomas, Megan C. "Isabelo de Los Reyes and the Philippine Contemporaries of *La Solidaridad*." In "The Book," 2, special issue, *Philippine Studies* 54, no. 3 (2006).

Thomas, Megan C. *Orientalists, Propagandists, and Ilustrados: Filipino Scholarship and the End of Spanish Colonialism*. Minneapolis: University of Minnesota Press, 2012.

Tilley, Helen. "Global Histories, Vernacular Science, and African Genealogies; or, Is the History of Science Ready for the World?" *Isis* 101, no. 1 (2010): 110–19.

Tiu, Macario D. *Davao 1890–1910: Conquest and Resistance in the Garden of the Gods*. Quezon City: UP Center for Integrative and Development Studies, 2003.

Tran, Anh Q. *Gods, Heroes, and Ancestors: An Interreligious Encounter in Eighteenth-Century Vietnam: Errors of the Three Religions*. AAR Religion in Translation Series. New York: Oxford University Press, 2018.

Tremml-Werner, Birgit. *Spain, China, and Japan in Manila, 1571–1644: Local Comparisons and Global Connections*. Amsterdam: Amsterdam University Press, 2015.

Urban, Greg. "Entextualization, Replication, and Power." In *Natural Histories of Discourse*, edited by Michael Silverstein and Greg Urban. Chicago: University of Chicago Press, 1996.

Vellema, Sietze, Saturnino M. Borras Jr., and Francisco Lara Jr. "The Agrarian Roots of Contemporary Violent Conflict in Mindanao, Southern Philippines." *Journal of Agrarian Change* 11, no. 3 (2011): 298–320.

Ventura, Theresa. "From Small Farms to Progressive Plantations: The Trajectory of Land Reform in the American Colonial Philippines, 1900–1916." *Agricultural History* 90, no. 4 (2016): 459–83.

Vergara, Benito M. *Displaying Filipinos: Photography and Colonialism in Early 20th-Century Philippines*. Quezon City: University of the Philippines Press, 1995.

Viatori, Maximilian Stefan, and Gloria Ushigua. "Speaking Sovereignty: Indigenous Languages and Self-Determination." *Wicazo Sa Review* 22, no. 2 (2007): 7–21.

Warren, James Francis. *The Sulu Zone, 1768–1898: The Dynamics of External Trade, Slavery, and Ethnicity in the Transformation of a Southeast Asian Maritime State*. Singapore: National University of Singapore Press, 1981.

Watson-Verran, Helen, and David Turnbull. "Science and Other Indigenous Knowledge Systems." In *Handbook of Science and Technology Studies*, edited by Sheila Jasanoff,

Gerald E. Markle, James C. Peterson, and Trevor J. Pinch, 115–39. Thousand Oaks, CA: SAGE, 2001.

Welch, Richard E. Jr. "Atrocities in the Philippines: The Indictment and the Response." *Pacific Historical Review* 43, no. 2 (1974).

Wernstedt, Frederick L., and Paul D. Simkins. "Migrations and the Settlement of Mindanao." *Journal of Asian Studies* 25, no. 1 (1965): 83–103.

Weston, Nathaniel Parker. *Specters of Germany: Colonial Rivalry and Scholarship in the Philippine Reform Movement and Revolution.* Quezon City: Ateneo de Manila University Press, 2021.

Wheatley, Jeffrey. "US Colonial Governance of Superstition and Fanaticism in the Philippines." *Method and Theory in the Study of Religion* 30, no. 1 (2018): 21–36.

Wheatley, Paul. "Geographical Notes on Some Commodities Involved in Sung Maritime Trade." *Journal of the Malayan Branch of the Royal Asiatic Society* 32, no. 2 (186) (1959): 3–140.

Whyte, Kyle Powys, Joseph P. Brewer II, and Jay T. Johnson. "Weaving Indigenous science, protocols and sustainability science." *Sustainability Science* (2015). http://link.springer.com/article/10.1007%2Fs11625-015-0296-6.

Wilkins, John S. *Species: A History of the Idea* Berkeley: University of California Press, 2009.

Wilson, Robert A., Matthew J. Barker, and Ingo Brigandt. "When Traditional Essentialism Fails: Biological Natural Kinds." *Philosophical Topics* 35, no. 1/2 (2007): 189–215.

Xhauflair, Hermine, Sheldon Jago-on, Timothy James Vitales, Dante Manipon, Noel Amano, John Rey Callado, Danilo Tandang, Céline Kerfant, Omar Choa, and Alfred Pawlik. "The Invisible Plant Technology of Prehistoric Southeast Asia: Indirect Evidence for Basket and Rope Making at Tabon Cave, Philippines, 39–33,000 Years Ago." *PLoS One* 18, no. 6 (2023).

Yalçinkaya, M. Alper. *Learned Patriots: Debating Science, State, and Society in the Nineteenth-Century Ottoman Empire.* Chicago: University of Chicago Press, 2015.

Yun, Kyoim. "Negotiating a Korean National Myth: Dialogic Interplay and Entextualization in an Ethnographic Encounter." *Journal of American Folklore* 124, no. 494 (2011): 295–317.

Yuval-Davis, Nira. *Gender and Nation.* Thousand Oaks, CA: SAGE, 1997.

Zaide, Gregorio F. *History of the Filipino People.* Manila: Modern Book, 1969.

Zaragoza, Ramon. *La Ilustración Filipina, 1891–1894.* Manila: RAMAZA, 1992.

Zarucchi, James L. "The Treatment of Aublet's Generic Names by His Contemporaries and by Present-Day Taxonomists." *Journal of the Arnold Arboretum* 65, no. 2 (1984): 215–42.

Index

Page locators with f indicate figures.

abacá: and cotton, 114, 126–27, 224n100; damage to crops, 35; dyes, 107, 119, 128; as export, 53, 111; large-scale production, 115–17, 116f, 121, 192; pakyaw, 222n54; publishing, 185; threads, 109, 120f, 122, 126f
Academia del Dibujo y Pintura (ADP), 63–67, 73, 214n46
achuete, 99
Adanson, Michel, 45, 48, 169
Ahern, George P., 142, 144, 190, 226n40
Aiton, William, 92–93; *Hortus kewensis*
Alcina, Francisco Ignacio, 67, 89–90
alliance imperialism, 112
Almansa, Alonso Cano, 64
American Museum of Natural History, 109, 201
animism, 149, 157
anitismo, 149
anonas, 99
Aquino, President Benigno, III, 194
areca nut, 108
Argüelles, Cayetano, 63, 66
asymptotic taxonomy, 24, 33, 51
Aublet, Jean-Baptiste-Christophe, 169

Bagobo: abacá, 111, 120, 122, 126–27; anthropological research, 25, 109, 118; Laura Watson Benedict, 110, 119–20, 123, 127, 151; changing community, 124–26, 130; elders, 128–29; identity, 111; Louisiana Purchase Exposition in St. Louis, 110, 119; Mount Apo, 118; Oleng, 108f; and settlers, 121; tapang, 123; tribal wards, 117, 122; use of synthetic fabrics, 126–28
Baker, John Gilbert, 70
Balangiga, 165, 179
balete, 3f, 4, 81, 151, 152f, 197
bamboo, 39f, 101, 108, 114
banana, 37, 111; fiber of, 19, 122, 192
Barad, Karen, 22, 110
Barcelona, Spain, 73, 80; exposition, 52, 213n101, 216n117
baroque style, 64–65
Bartlett, Harley Harris, 164
Bataan, 76, 117, 136f, 138
Beccari, Odoardo, 78–80, 187
Beille, Lucien, 177–78
Benedict, Laura Watson: background, 110, 119; fieldwork, 119–23, 125, 127–28, 151; methods of acquiring textiles, 129–31. *See also* Bagobo
Bentham, George, 48, 68, 187
Berlin, Brent, 100
Berlin Conference, 189, 212n79
betel nut, 50, 108
Bignoniaceae family, 75
Bignoniáceas, 75f
bihod, 193
binayayo, 50, 212n98
binomials: and common names, 99, 161–62, 165, 169–72, 182, 195; Latin, 14, 26, 52, 159
Bisaya, 89, 149

Blanco, Francisco Manuel, 229n44; background, 30, 37; *Flora de Filipinas*, 41, 55, 57, 67, 72, 91, 161; observations of women's labor, 38; scholarly dialogue, 78, 93, 164–65, 169, 190
Bleichmar, Daniela, 73, 76, 215n86, 216n90
Bolívar y Urrutia, Ignacio, 69f
bolohan, 37
Bonaparte, Napoleon, 36
bontot: cabayo, 37; pusa, 50
Breedlove, Dennis, 100
Brown, William Henry, 114–15, 121, 128, 190
Brussels Conference Act of 1890, 189
bulacan, 50
bulao, 50
bunlay, 50
Bureau of Government Laboratories, later renamed the Bureau of Science, 113; Paul C. Freer, 114, 141, 148; Dean C. Worcester, 113, 151
Burman, Nicolaas Laurens, 161
buso, 151, 153

cabog, 37
Calderon, Fernando R., 149
Calle, Blas Sierra de la, 102
Canarium ahernianum Merr., 144, 145f
Candolle, Alphonse Pyramus de, 47, 162
Candolle, Augustin Pyramus de, 40, 47, 52, 170
Cañizares-Esguerra, Jorge, 24, 188
cargadores, 147f
casampagahan, 91
Cebuano, 87, 193
Chicago World's Fair (1893), 104
Chofré, Salvador, 72, 87
Cicca acidissima, 178
Clemens, Joseph, 165–67, 174–75, 179, 185
Clemens, Mary Strong, 168f; and João de Loureiro, 167–72, 177, 181, 187; education and training, 159–60, 165–67; fieldnotes, 153; and Elmer D. Merrill, 25, 158, 169, 174–81; nomenclature, 161–62
clerics, 89–90
co mè, 178
cognates, 181
Colenso, William, 170
collaboration, 138, 158, 164, 173, 175

collectors (of plants): Anglo-European, 178; Bagobo communities and, 118; Bureau of Science, 159, 185; Gregorio Edaño, 166–67; Carl Linnaeus, 12; Elmer D. Merrill and, 161, 165, 174–75; militarization and, 117; Mount Apo, 131; Nvlvk'ö, 153; Philippine Exposition Board, 119; Philippine National Herbarium, 154; Maximo Ramos, 144, 166–67; role of, 138–40; Cornelis G. G. van Steenis, 188; women, 18
College of Agriculture in Los Baños, 185
Colmeiro, Miguel, 35–36, 48, 170
colonial terrain, 5, 207n45
colonization, 2; in Africa, 189; "Filipinization," 103, 186; France in Indochina, 179; Spanish, 4–5, 7, 31; US, 103, 139, 142, 179
Comisión de la Flora y Estadística Forestal de Filipinas, 57, 93, 185
Comisión de Ultramar, 62, 68, 80
corolla, 76, 99, 177
Costa i Cuxart, Antoni Cebrià, 36
Crassula pinnata, 178
Creole, 5, 17–18, 43, 73, 87, 95
Cuba, 35–36, 59, 62, 165, 205n8; War of Independence, 112
Cuéllar y Villanueba, Juan José Ruperto de, 67, 209n8, 218n36

Delgado, Juan José, 210n29
Díaz Pascual, Concha, 64
dinulung, 193
Doeppers, Daniel F., 39–40
Dolichandrone rheedh, 76
Domingo, Damián, 64, 101. *See also* tipos del país genre
Doñamayor y Moreno, Celedonio, 142
drawings, 23, 63, 65, 77, 100, 139; instruments, 73
Dupré, John, 51

Edaño, Gregorio, 154–56, 155f, 166. *See also* Maximo Ramos
Elmer, Adolph D. E., 117
Emmerson, Donald K., 188
England, 50, 78, 171
Escuela de Agricultura, 34, 63, 186, 191
Escuela de Bellas Artes de San Fernando, 43, 64, 73

Escuela Náutica, 63, 214n44
Espejo y Culebra, Zoilo, 35
Euforbiáceas, 71f
Eurycoma longifolia, 178
Everett, Harry D., 143
Exposición de las Filipinas in Madrid (1887), 52, 80
Exposición Universal in Barcelona (1888), 52, 80
Exposition Universelle in Paris (1889), 52

femininity, 86, 95–96, 102, 105
field assistants, 138, 154–57
field guides, 23, 25, 139–40; labor disputes, 153–54, 157, 161; recruitment of, 156
field labor, 25, 135–40, 143, 154–57, 175
Filipino: artist trope, 216n91; cargadores, 147; field assistants, 138, 154–57; forest rangers, 143; identity, 5, 18; nationality, 104; plant collectors, 141; poetry, 98–99; political definition, 77, 216n91; settler colonialism, 117
Filipino Enlightenment, 5, 57, 191
First Philippine Republic, 113, 190
Florentino, Leona Josefa, 96–99, 201
Flores, Patrick D., 64, 216n91
Folgueras, Mariano Fernández de, 209n8
France, 36, 47, 179
Freer, Paul C., 114, 141, 148
French: botanists, 45, 142, 169, 177–78, 207n45; Candollian system, 40; empire, 13, 111, 169, 179; medico, 178–79; Revolution, 20; scholarship, 17, 173; *soverain*, 19; Spanish War of Independence, 36; superstition, 157; vernacular and translation, 5, 169, 174, 177, 181; work with Carl Linnaeus, 12. *See also* Indochina

García, Regino, 17, 33, 55–56, 65, 88, 118, 164; Inspección General de Montes, 190; Jardín Botánico de Manila, 23, 37, 40–44, 57, 63, 185, 189; *Sinopsis*, 57–58, 68–69, 76–77, 93, 187, 214n27
Gates, Frank Caleb, 147
gender: ideology, 14, 102–3; labor, 38; sampaguita, 86, 89, 97, 99; vernaculars, 17–19
German: botanists, 48, 78, 81, 113, 117, 142, 214n27; explorers, 139; *Flora Malesiana*, 187–88; language for scholarship, 17, 61, 173

Germany, 80, 112; nationalism, 7, 20, 61, 230n71; Nazi, 173
Great Depression, 173–74
Guerrero, León María, 191
guyod, 37

Harrison, Francis Burton, 104
hemp, 111, 119, 121, 126f, 127
Hooker, Joseph Dalton, 48, 68, 187
Hooker, William Jackson, 70
Houttuyn, Maarten, 161

Iberia, 78
ichthyology, 188
ikat, 108f, 128
ilang-ilang, 98, 104, 220n85
Iloilo, 76
La Ilustración filipina, 101–2, 226n35
ilustrado production, 77, 191
Indochina, 5–6, 160, 169, 174–76, 180–81, 231n87; France's colonization, 7, 167, 179
Inspección General de Montes, 59, 67, 78, 81; Filipino Enlightenment, 191; Regino García, 190; publishing, 60, 62, 66, 71, 76, 82, 185; Sebastián Vidal, 60
International Botanical Congress: development of, 14, 24, 46–48; other governing bodies, 20, 23, 31, 33, 189; upholding Latin, 17, 52, 174; Elmer D. Merrill, 25–26, 161–63, 171–72, 187
International Union of Botanical Sciences, 173
Isabella II, 36, 48, 79
"Las Islas Filipinas," 5, 60
itlog balang, 50
Izquierdo, Rafael de, 79

Jack, William, 142, 225n25
Jackson y Zaragoza, Emina, 67
Jagor, Fedor, 60–61, 139, 186, 226n46
Jardín Botánico de Manila, 50–51, 66–67, 185–86; Odoardo Beccari, 78; Miguel Colmeiro, 35–36; Vicente Cutanda, 35; Escuela de Agricultura, 34, 63, 186, 191; Zoilo Espejo, 35, 60; Regino García, 23, 40–44, 57, 63, 189–90, 193; leadership, 82, 141; origin of, 4, 23, 34, 52; records, 191, 232n17; rice, 31, 37, 41, 211n54, 213n101

jasmine, 87, 90–93, 98, 101; cultural production, 219n59. *See also* Francisco Ignacio Alcina; sampaga
Joaquin, Nick, 216n91
Jordana y Morera, Ramón, 71

Kamel, Georg Joseph, 66, 91
Kern, Johan Hendrik Caspar, 101

laborers, 17; Chinese, 180; dichotomy with scientists, 141, 143 Filipino, 25, 82, 118, 121, 135
Lapeyrouse, Isidore Picot de, 142
Latin: binomials, 14, 26, 104, 161–62, 165, 171–72; *J. sambac*, 93; Jardín Botánico de Manila, 52; Carl Linnaeus, 12; Elmer D. Merrill, 163, 181; nomenclatural conventions, 41, 43, 47, 99, 109, 115, 161; *Phyllanthus distichus*, 178; rice, 23; sampaguita, 88, 91; *super*, 19; superstition, 157; *varietās*, 44; *verna*, 15; and vernacular names, 2, 16–17, 51, 169–71
Latin America, 5, 24; racial categorizations, 18; US influence in, 112; wars of independence, 36
Latin babble, 26, 162, 181
Lázaro e Ibiza, Blas, 36
League of Nations, 189
Lecomte, Paul Henri, 78
Linnaeus, Carl, 9, 12; cryptogamia, 207n42; definitions, 211n65; Linnaean systematization, 12, 13, 16, 50–52, 189; *N. sambac*, 92–93; *Philosophia botanica*, 44–45; *Species plantarum*, 50, 218n34; *Systema naturae*, 92
lithograph plates, 58, 71
Llanos, Antonio, 165
location: mentioned in labels of plants, 4, 160; of the Philippines, 5, 7, 21; significance of, 30, 193
Lodén, José, 67, 77
López y Jaena, Graciano, 99
Loureiro, João de, 25, 160; *Flora cochinchinensis*, 161–62, 167–72, 177, 181, 187
Lozano, Jose Honorato, 39f
Lumbera, Bienvenido L., 95, 98, 218n27
Luna y Novicio, Juan, 43

Macaranga mappa, 71
MacArthur, General Arthur, 190

Madrid, Spain: colonial knowledge production, 24, 36, 57, 67, 78, 191, 209n8; Exposición de Filipinas, 52, 80, 216n117; Regino García, 40, 43; publishing, 71, 99, 141. *See also* Jardín Botánico de Manila
Malcampo y Monge, José, 61–62
Malesia, 59, 161; Odoardo Beccari, 78, 187; collaborative research in, 61, 80–82, 173; *Flora Malesiana*, 187–88; geography of, 77–78
Mallotus moluccanus Müll., 70
Malolos Constitution, 88, 113
Manila: aristocratic, 58, 63–64, 73, 103; botanical garden, 209n8; Mary Strong Clemens, 166–67; commercialization and urbanization, 59–60; destruction of, 170; Ramon García, 43–44, 52; gender portrayals, 101–2; "Jocelynang Baliwag," 96; Elmer D. Merrill, 179, 190; pensionado programs, 186; Philippine-American War, 113; publishing, 86–87; rice, 37, 39–41, 52; sampaguita, 24, 89–91, 95–97, 193; University of Santo Tomas, 79; Sebastián Vidal, 61–62, 67, 80–81, 93. *See also* Jardín Botánico de Manila
Maravall, José Antonio, 64
Martínez, Juan Antonio, 209n8
Martinez y Lorenzo, Felix, 101, 102f
Martínez Vigil, Ramon, 164
Menonville, Nicolas-Joseph Thiéry de, 207n45
Merrill, Elmer D.: cataloging style, 115; and Mary Strong Clemens, 25, 158, 169, 174–81; on collaboration, 138, 161, 164, 173, 181; field assistants, 164; International Botanical Congress, 25–26, 161–63, 171–72, 187; Latin, 163, 181; Manila, 179, 190; New York Botanical Garden, 156; nomenclature, 160–61, 169–74, 178, 181, 187; and Maximo Ramos, 154; sent to Philippines, 114; Southeast Asia, 160; and Sebastián Vidal, 213n7
mestizo: dominance in scientific writing, 5, 17–18; Damián Domingo y Gabor, 64; Ramon García, 43, 63, 73
Mindanao: abundant plant life, 116–17, 121; Laura Watson Benedict, 110, 119; colonization of, 131, 166; Davao Gulf, 109; ilang-ilang, 194, 220n85; missionaries in, 179; Moro fighters, 113; population, 124–25; Maximo Ramos, 141, 154; weavings, 192

Ministerio de Ultramar, 34–35, 60, 62, 78–79
missionaries, 17, 37, 67, 90; Francisco Manuel Blanco, 30; Catholic, 37, 66, 148; Methodist, 167, 179; Protestant, 179
Mojares, Resil B., 57, 87, 99, 104
Molina, Antonio, 86
Murillo, Bartolomé Esteban, 64
Murphy, Frank, 104
Musa textilis, 107, 114, 120
Mutis, José Celestino, 76–77. *See also* Daniela Bleichmar

National Museum of the Philippines, 194
Nazario, Tomás, 67, 77
Née, Luis, 120
Nepomuceno, Juan Serapio T., 32f, 49f
New Spain, 5, 207n45, 227n79
New York, 104, 109, 163, 165, 172
New York Botanical Garden, 117, 156, 169, 174
Noceda, Juan de, 90, 178
nomenclature or nomenclatural, 95, 99; Michel Adanson, 169; Alphonse Pyramus de Candolle, 47–48; Mary Strong Clemens, 160, 162–63, 174; William Colenso, 170–71; and colonialism, 5; fixity, 179, 182, 195–96; Regino García, 17, 51; Latin babble, 25–26, 41, 43, 162; local, 18; Elmer D. Merrill, 160–61, 169–74, 178, 181, 187; standards of, 14, 16; Trinidad Pardo de Tavera, 99; variation across the Philippines, 89, 95; vernacular, 16–17, 26, 160, 178
"Non-Christian," 109, 117, 125
Norzagaray, Fernándo, 34
Nueva Ecija, 52, 76
Nyctanthes sambac, or *N. sambac*, 91

obreros alumnos, 63, 186
Ocimum americanum, 74f
Oldenlandia corymbose, 178
Oleng, 107, 108f, 109, 130, 192
"once known," 116, 136; in relation to "unknown," x
Order of Saint Augustine in the Philippines, 73
Oryza sativa, or *O. sativa*, 23, 33, 42f, 190; catalog and nomenclature, 30, 41, 51, 233n31

Pagsulong sa Karunungan, 150
pagtuotuo, 149, 157
pamahiin, 140, 149–50, 157

Pangasinan, 39–40, 52
Pardo de Tavera, Trinidad H., 43, 55, 57, 99, 191; Ramon García, 77; *Plantas medicinales de Filipinas (Medicinal Plants of the Philippines)*, 99; superstitions, 149
Paris Peace Conference, 189
Pastor-Roces, Marian, 210n41
Paterno, Pedro, 57; *Nínay*, 149; Pact of Biak-na-Bato, 103; *Sampaguitas*, 96
Paterno y Ignacio, Dolores, 96
Patnubay ng̃ Bayan, 150
pensionados, 34, 64, 185
pest blights, 192
Pharnaceum incanum, 178
Philadelphia, Pennsylvania, 62, 80
Philippine–American War, 25, 113, 163, 165, 190
Philippine Exposition Board, 119
Philippine Journal of Science, 185, 137f
Philippines: anitismo, 149; botanical drawings, 18, 63–64, 66, 77, 82; collaboration in, 158, 164; colonial occupation of, 4–8, 21, 31, 36; environment, 1–2; environmental spirits, 150; everyday experiences, 101, 103; field assistants, 131, 139, 142–43, 147, 151, 153; First Republic, 88, 190; General Hospital, 149; geography, 6f, 71, 137f; ilustrados, 86; intellectuals, 87; "Jocelynang Baliwag," 85–87, 96–97; languages of, 17, 90, 95, 170, 184; lawmakers, 194; Malaspina Expedition, 67, 120; missionaries in, 89; Moro Province, 117; Mount Apo, 109, 117–19, 131, 153; national flower, 87, 97; nationhood, 9, 103–4; pamahiin, 150; Revolution of 1896, 5–7; rice varietals, 30, 50, 52–53, 193; settler colonialism, 125; sexism, 18; sovereignty, 20; and US Bureau of Science, 141, 185; University of, 144; wars, 112–13; weaving, 25, 108–9, 115, 125; womanhood, 98, 102. *See also* abacá; Bagobo; Mindanao
Phyllanthus distichus, 178
Pitard, Charles-Joseph Marie, 178
Plano del Jardín Botánico de Manila (Plan of the JBM), 76–78, 82, 185–87, 190–93
Puerto Rico, 36, 59, 62

Rafael, Vicente L., 148, 193, 196
Ramos, Maximo, 141–48, 166, 180, 190–91; death, 154–59

Raven, Peter H., 100
Real Jardín Botánico, 36, 40, 44, 48
Real Sociedad Económica, 34, 63
Renacimiento Filipino, 1, 183
Reni, Guido, 64
Retana, Wenceslao E., 149
revolution: American, 111; anticolonial, 7, 82, 85–86, 88; French, 20; Glorious, 36, 48; insurgents, 20; Pact of Biak-na-Bato, 103; Philippine, 5, 79, 104, 112–13, 170; political, 15, 194; Scientific, 122; Spanish, 112–13; and superstition, 150
Reyes, Isabelo de los, 57, 135; *El Folklore filipino*, 72, 97–98, 149
Reyes, Miguel de los, 67, 77, 98–101
Reyes, Raquel A. G., 67, 102–3, 220n76
Ribera, Jusepe de, 64
rice: Filomeno V. Aguilar, 124; Francisco Manuel Blanco, 30, 37; Exposición de Filipinas, 81; Regino García, 33, 43–44, 51–52, 80, 185, 189, 213n101; gendered labor, 210n39; genomics research, 233n42; global history, 30; harvest, 49f; husking, 32f, 38f, 50, 210n41, 211n54; Jardín Botánico de Manila, 23, 31, 37, 41, 211n54, 213n101; imports and exports, 53; increased production of, 38–40; mortar and pestle, 32f; *Oryza sativa*, 24, 51; pounding of; puto bumbong, 39f; scientific naming of, 50; seed catalogs, 33, 41; sensory qualities of, 29; Spanish colonization, 31; taxonomies; textures of, 19; varietals' names, 37, 41, 43–44, 193, 210n29
Rizal, José, 43, 57, 214n22; Exposición de Filipinas, 81; Regino García, 99; nationalism and, 193; *Noli me tangere*, 20, 35
Roberts, Nathaniel, 191
Rocha é Icaza, Lorenzo, 73, 79
Rodriguez, Eulogio B., 104
Rumphius, Georg Eberhard, 92–93, 161, 169

Samar, 6, 115, 165–66, 179
sampaga, 85–87; Francisco Ignacio Alcina, 89–90; Manuel Blanco, 91, 93; distinguished from sampaguita, 95–96, 98–99; gendered portrayal, 99; Johan Hendrik Caspar Kern, 101; Leona Josefa Florentino, 97–98; Georg Joseph Kamel, 91; missionaries; naming, 90; Pedro Paterno, 103; in poetry, 90; as political project, 104; variations of naming and wording, 91; varieties of, 192
sampaguita, 87; distinguished from sampaga, 95–96, 98–99; Latin name, 88; Félix Martínez y Lorenzo, 101, 102f; Trinidad H. Pardo de Tavera, 100–101; Pedro Paterno, 103; as political project, 89, 93–95, 97, 104, 163, 192–94; in song, 219n50, 219n59
Sanlucar, Pedro de , 90, 178
Sanskrit, 101, 162
Santa Cruz, Davao, 109, 117, 119, 194
Santiago, Luciano P. R., 64, 73
Sauropus androgynus, 176f
scientific statecraft, 24, 57–58, 62, 187
seeds, 30 38, 40, 71, 119, 126f
Seetzen, Ulrich Jasper, 142
Semper, Carl, 139
sensory, 24, 29
shirtwaist, 127f
Shizhen, Li, 161
Sirks, Marius Jacob, 172
Smith, US general Jacob H., 165
Smith, Pamela H., 122
Solander, Daniel Carlsson, 93
Southeast Asia, 160; Mary Strong Clemens, 153; development of the term, 6f, 188, 232n15; discourse about, 188; Elmer D. Merrill, 160; trading of plants from, 11, 30; vernacular science, 16
sovereign vernaculars: imperial dominance of naming, 15–19, 21–23; local relationships with plants, 88, 140, 191, 195
Spain: baroque style, 64; botany in the Philippines, 23–24, 35–36, 40, 80, 82, 185; colonial influence, 5, 7, 25, 116–17; competition with Germany, 61, 80; Ramon García, 43–44, 80; Isabella II, 48; Ministerio de Ultramar, 34; New, 5, 207n45, 227n79; Philip II, 5; profitability from colonies, 34, 59, 67, 187, 191; synopses, 69; Sebastián Vidal, 62, 78, 80. *See also* International Botanical Congress; Jardín Botánico de Manila
Spanish: academic tradition, 64; awards to researchers, 80; botany and colonialism, 2, 14, 17–19, 57; cereals, 44; churches, 79; College of Agriculture students, 146f, 147; colonial government, 103; Cuban War of Independence, 112; documentation of

the Bagobo, 118; "gentil," 96; language for publishing, 17, 60–62, 72, 86, 97–98; libraries, 78; mandates by the Crown, 36, 63, 79, 88–89, 116; missionaries, 89; monarchy, 185; multiraciality, 43; names, 41; poetry, 95; racial regime, 205n8; Real Compañía de Filipinas, 209n8; scientists, 21; Society of National History, 48; state secrecy, 40, 67; superstition, 140, 157; territories, 105, 111–12; trade, 121; War of Independence, 36

Spanish–American War, 7, 112–13, 165

stamens, 76, 177

state flower, 104

Stereospermum quadripinnatum, 75–76

Subramaniam, Banu, 48, 51

superstition, 22, 150, 157; and acts of resistance, 25, 153, 157–58, 195; discourse on, 140–41, 148–50; effect on laborers, 138, 151–52; knowledge systems and, 192, 198; significance of the word, 138

Sydow, Paul, 175

synonymy, 162, 170, 181, 195

Tagalog: rice production, 32f, 38, 212n98; superstitions, 150, 157; translations, 37, 43, 50, 87, 91, 95, 218n36, 219n50

Tanay, 76

tapang, 123–25, 224n100

Tarlac, 52, 60, 76

taxonomy, 16, 99, 108, 172–73; asymptotic, 24, 33, 51–52, 89

terra ignota, 59

Thomas, Jerome B., Jr., 103–4

Thomas, Megan C., 18, 149, 199, 203

Tiaong, 193

tipos del país genre, 65. *See also* Damián Domingo

Torre, Carlos María de la, 79

transformations: colonial, 109, 111; woven, 192

translators, 25, 90, 139, 167

Treaty of Paris, 7

tuai, 3–4, 151–52

van Rheede, Hendrik, 92–93, 161, 218n36

van Steenis, Cornelis G. G., 155f, 187–88. *See also Flora Malesiana*

Verdoorn, Frans, 173

Vidal, Sebastián: Odoardo Beccari, 78; collaboration with Aeta, 226n35; death, 81; Exposición de Filipinas, 81; fieldwork and expeditions, 118, 143; and Ramon García, 58, 72, 80, 213n6; Inspección General de Montes, 60, 67; leadership of, 24, 62; and Elmer D. Merrill, 213n7; naming plants, 94f, 164; *Reisen in den Philippinen*, 60; and Jose Rizal, 214n22; scientific training and relationships, 61, 68, 187; *Sinopsis*, 57–58, 68–69, 76–77, 93, 187, 214n27; *La Solidaridad*, 220n72

Vidal y Soler, Domingo, 67, 185

Visayas, 6f, 60, 67, 121, 125, 210n39; natural history of, 89, 116, 149, 193. *See also* Francisco Ignacio Alcina

Vocabulario de la lengua tagala, 90, 178

von Humboldt, Alexander, 187

waling-waling, 194

Warburg, Otto, 117

Weston, Nathaniel Parker, 61, 226n46

World War I, 166, 189, 230n71

World War II, 7, 188–89

Zambales, 39, 52

Zollinger, Heinrich, 61